Engineering Materials

More information about this series at http://www.springer.com/series/4288

Asish K. Kundu

Magnetic Perovskites

Synthesis, Structure and Physical Properties

Springer

Asish K. Kundu
Indian Institute of Information Technology,
Design and Manufacturing Jabalpur
Jabalpur, Madhya Pradesh
India

ISSN 1612-1317 ISSN 1868-1212 (electronic)
Engineering Materials
ISBN 978-81-322-2759-5 ISBN 978-81-322-2761-8 (eBook)
DOI 10.1007/978-81-322-2761-8

Library of Congress Control Number: 2015960829

Printed on acid-free paper

This Springer imprint is published by SpringerNature
The registered company is Springer (India) Pvt. Ltd.

Dedicated to Profs. C.N.R. Rao and B. Raveau

Preface

The field of magnetic perovskites is attracting increasing importance from the condensed matter physicists and materials science community in the last few years because of their potential applications in storage and sensing/actuating devices. This book demonstrates the experimental results on the magnetic, electronic, and multiferroic properties of manganese, iron, cobalt, nickel, and bismuth centered rare earth perovskites. It is organized into four chapters. Chapter 1 provides a brief introduction of various interesting phenomena in magnetic perovskites, e.g., colossal magnetoresistance, electronic phase separation, and multiferroic properties of the general formula ABO_3.

Chapter 2 describes the results of the investigations on electronic phase separation and glassy ferromagnetism of the hole-doped perovskite manganites and cobaltites. Measurements of magnetic and electron transport properties have been discussed for different types of perovskites. The various aspects studied include the effects of A-site cation radius and the novel effects of cation size disorder. Similarly, the ordered and disordered effects in perovskite structure and related aspects in hole-doped perovskite cobaltites are described in Chap. 3.

Finally, in Chap. 4 we have discussed the bismuth based (ferro-)magnetic perovskite, which shows multifunctional behavior. As for the present trends toward device miniaturization and high-quality data storage, an integration of multifunction into one material system has become highly desirable. The various multiferroics discussed in this book represent one such type of perovskite materials, which do offer the opportunity for humans to develop an efficient control of magnetization and/or polarization by electric field and/or magnetic field, as represented in Chaps. 2 and 3, and to explore their multi-implications.

Contents

About the Author

Asish K. Kundu is Associate Professor in the Discipline of Physics at IIITDM-Jabalpur, India. He received his master's degree in Physics from BHU, Varanasi in 2001, and obtained his Ph.D. in Magnetic Materials from JNCASR, Bangalore in 2006. He was head of Natural Sciences Discipline at the IIITDM-Jabalpur from 2012–2014, where he presently continues his research and teaching. His research in perovskites includes magnetic materials, magnetoresistance materials, and multiferroicity of transition metal oxides. He has published several research articles and two books in perovskite materials. He was a CNRS research fellow at CRISMAT Laboratory, France (2006–2008), an Indo-Sweden visiting scholar at Angstrom Laboratory, Sweden (2004 & 2005) and a DRDO scientist at Solid State Physics Laboratory, Delhi (2008–2009). He was also a recipient of UGC CSIR (NET) Junior & Senior Research Fellowship for the year 2001–2006 and awarded BOYSCAST Fellowship for the year 2011 by DST Government of India.

Chapter 1
Introduction to Magnetic Perovskites

1.1 Importance of Perovskites

Perovskites constitute one of the most fascinating classes of solid materials, and show a very wide variety of physical phenomena and properties. There has been extensive research on ABO_3-type perovskites of the general formula $Ln_{1-x}A_xMO_3$ (Ln = rare earth, A = alkaline earth or bismuth, M = transition metal oxides). Some of the novel properties of the perovskites are known for some time, particularly the paramagnetic (PM) to ferromagnetic (FM) transition at Curie temperature (T_C) and the associated insulator–metal transition (T_{IM}) in the case of manganites and cobaltites [1]. The discovery of colossal magnetoresistance (CMR) in doped manganites has renewed great interest in this system since the early 90s. Apart from CMR, the perovskite manganites exhibit rich phase diagram spanning a wide range of magnetic properties and phenomena like charge ordering (CO), orbital ordering, electronic phase separation (EPS), and spin glass behavior [2, 3]. These phenomena represent a combined interaction between the spin, the lattice, the charge, and the orbital degrees of freedom. Such interactions are manifested in single crystal and polycrystalline samples as well as in the thin films. Experimentally, it has been observed that the doped perovskite manganites, $Ln_{1-x}A_xMnO_3$, reveal rich phase diagrams which include spin, charge, and orbital ordered phases as well as some magnetic phases. The properties of these perovskites could be tuned either by external factors or by chemical means. In certain critical range of cation doping at the A-site, the perovskites exhibit simultaneous occurrence of ferromagnetism and metallicity, along with a CMR in the vicinity of T_C or T_{IM} [4]. Similarly, the cobalt perovskites, $Ln_{1-x}A_xCoO_3$, are also interesting in which they exhibit compositionally controlled insulator–metal transitions and ferromagnetism, the ferromagnetic phase being metallic [1, 5]. The properties of both the perovskite manganites and cobaltites are affected by the size of the A-site cations.

Some of the perovskite-centered transition metal oxides (TMO) are known to exhibit compositional and electronic inhomogeneities arising from the existence of

A.K. Kundu, *Magnetic Perovskites*, Engineering Materials,
DOI 10.1007/978-81-322-2761-8_1

more than one phase in crystals of nominally monophasic composition. This is understood in terms of electronic phase separation described in the literature [3]. Such a phenomenon occurs because of the comparable free energies of the different phases [2, 3]. The phase-separated hole-rich and hole-poor regions give rise to anomalous properties such as weak FM moments in an antiferromagnetic (AFM) regime. A variety of magnetic and electronic properties manifest themselves in $Ln_{1-x}A_xMnO_3$ depending on the various factors such as the A-site cation size and size disorder as well as external factors such as temperature, magnetic field, etc. In the last few years electronic phase separation in perovskite manganites and cobaltites has attracted considerable attention [2, 3].

The interesting properties of TMO-based perovskites are due to unique nature of the outer *d*-electrons, the metal–oxygen bond varying anywhere from ionic to metallic. The phenomenal range of electronic and magnetic properties exhibited by TMO is equally interesting. There are oxides with metallic properties (e.g., RuO_2, ReO_3, $LaNiO_3$) at one end and with insulating behavior ($BaTiO_3$) at the other end [5]. There are oxides that traverse both these regimes with the change in temperature, pressure, or composition (V_2O_3, $La_{1-x}Sr_xVO_3$). Interesting electronic properties also arise from charge density waves ($K_{0.3}MoO_3$), charge ordering (Fe_3O_4), and defect ordering ($Ca_2Mn_2O_5$, $Ca_2Fe_2O_5$). Examples of TMO are known with diverse magnetic properties, like ferromagnetic (CrO_2, $La_{0.5}Sr_{0.5}MnO_3$), ferrimagnetic (Fe_3O_4, $MnFe_2O_4$), and antiferromagnetic (NiO, $LaCrO_3$). Many oxides possess switchable orientation states as in ferroelectric ($BaTiO_3$, $KNbO_3$) and ferroelastic ($Gd_2Mo_3O_{12}$) materials. The discovery of high-temperature superconductivity in cuprates has focused worldwide interest on the physics and chemistry of TMO since 1986. The unusual properties of TMO that distinguish them from metallic elements, covalent semiconductors, and ionic insulators arise from several factors [5]. One of the reasons is that the oxides of *d*-block transition elements have narrow electronic bands, because of the small overlap between the metal *d* and oxygen *p* orbitals. The band widths are typically of the order of 1 or 2 eV (rather than 5–15 eV, for metal).

More specifically, in the TMO strong interplay between lattice, charge, spin, and orbital degrees of freedom provides a fantastic playground to tune their physical properties. Multifunctional materials (multiferroics, spintronics, etc.) have attracted increasing attention due to their possible applications toward storage materials and fundamental physics [6–9]. Among the naturally existing oxides, the presence of both ferromagnetism and ferroelectricity is a rare phenomenon, due to incompatibility between magnetism and ferroelectricity. This incongruity could be at the origin of a limited number of multiferroics, though the researchers are looking for such materials from more than six decades. This phenomenon often also occurs in TMO centered perovskites having the general formula ABO_3. In the process of exploration of a multiferroic perovskite the following facts are now well established: (i) ferromagnetic (FM) and ferroelectric (FE) behaviors are mutually exclusive due to the d^0 electronic structure of the B-site element, (ii) the occupation of different B-site cations with varying ionic radius provides an opportunity to realize a polar ground state, and (iii) the lattice distortion induced by cations with lone pair electrons such as Pb^{2+} or Bi^{3+} plays a primordial role on the FE properties as reported for $PbTiO_3$ in comparison with $BaTiO_3$ [6]. The most well-known examples of existing perovskite multiferroics are $BiFeO_3$ and $BiMnO_3$ [6–9]. In the recent years a number of perovskites (e.g., $La_2Mn(Co/Ni)O_6$, Bi_2MnNiO_6, $LnMnO_3$, $LnMn_2O_5$, $LnCrO_3$, $YBaCuFeO_5$, CuO, $MnWO_4$, $(LnGa)FeO_3$, $LnFe_2O_4$, Ln_2BaNiO_5 etc.) have been reported to exhibit simultaneous electrical and magnetic ordering [6–9].

1.2 Description of Perovskite Structure

The stoichiometry perovskite of the general ABO_3 form a large group of transition metal oxides whose structure is based on that of ReO_3 [5]. Stoichiometry perovskite is a simple cubic structure (Pm-3m) as shown in Fig. 1.1. However, many perovskites distort a little from this structure even at room temperature.

The perovskite structure is most stable when the Goldschmidt tolerance factor, t, is unity (for cubic structure), which is defined by $t = (r_A + r_O)/\sqrt{2}\,(r_B + r_O)$ where, r_A, r_B, and r_O are the average ionic radius of the A, B, and O ions, respectively. Deviation of 't' from unity leads to the structural distortion. For a small deviation in t (i.e., $t < 1$), the crystal structure changes from cubic to rhombohedral or orthorhombic symmetry. In this situation the ⟨Mn–O–Mn⟩ bond angle decreases from 180°. The perovskite structure occurs only within the range $0.75 \leq t \leq 1.00$. To stabilize the A- and B-site cations in their respective 12- and sixfold coordinations, the lower limits of their radius should be set as $r_A > 0.90$ Å and $r_B > 0.51$ Å. The stability of the perovskite structure of manganites and cobaltites

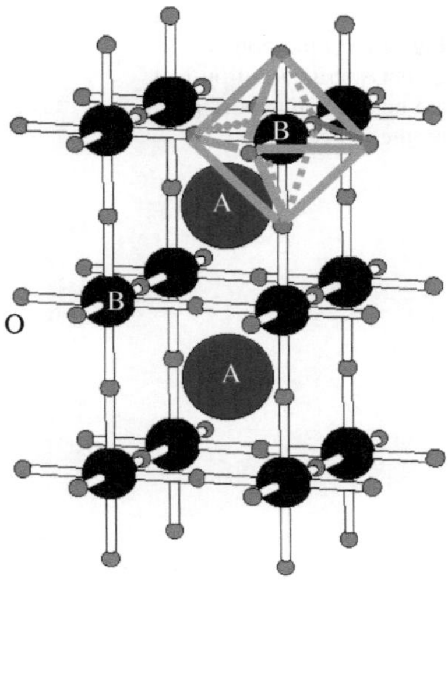

Fig. 1.1 Stoichiometric perovskite structure ABO_3

depends on the relative size of the Ln/A and M ions in $Ln_{1-x}A_xMO_3$. In rare-earth manganites Ln/A cation is surrounded by eight corner-sharing MnO_6 octahedra, which build a 3D network. The smaller ionic radius of the cations results in a lower value of 't,' and consequently more is the lattice distortion. The increase in lattice distortion significantly decreases the ⟨Mn–O–Mn⟩ bond angle from 180°, which strongly affects the physical properties of perovskite manganites. When $t < 1$, there is a compression of the Mn–O bonds, which in turn induces a tension on Ln–O bonds. A cooperative rotation of the MnO_6 octahedra and a distortion of the cubic structure counteract these stresses. For $0.75 < t < 0.90$, the MnO_6 octahedra tilts cooperatively to give an enlarged orthorhombic (Pbnm) structure of $GdFeO_3$ (Fig. 1.2). For $0.90 < t < 1.0$, buckling of octahedra is not found and small distortion leads to lower symmetry structure. Hence, tilting of MnO_6 octahedra along [111] direction gives rise to the rhombohedral structure ($LaAlO_3$), whereas along [001] direction leads to tetragonal structure ($SrTiO_3$). The perovskite oxides can be crystallized not only in orthorhombic and cubic structure, but in tetragonal, hexagonal, rhombohedral, and monoclinic structures as well. The orthorhombic structure has four formula units per unit cell ($Z = 4$), the rhombohedral has $Z = 2$, and hexagonal has $Z = 6$ as compared to one in the case of ideal cubic perovskite.

It is noteworthy to mention that the TMO constitute one of the most interesting classes of materials exhibiting a wide variety of structures and physical properties. TMO crystallize in a variety of structures, and bonding in these materials can vary from ionic (e.g., MgO, $Fe_{1-x}O$) to metallic (TiO, ReO_3). TMO possessing several types of complex structure have been characterized in last several years [2–5]. These include not only the well-known perovskite, spinel, pyrochlore, and hexagonal structures, but also the octahedral tunnel, tubular, and lamellar structures. Many of the TMO are not truly three dimensional, but have low-dimensional features. For example, $Ca_3Co_2O_6$ with a quasi-one dimensional structure and

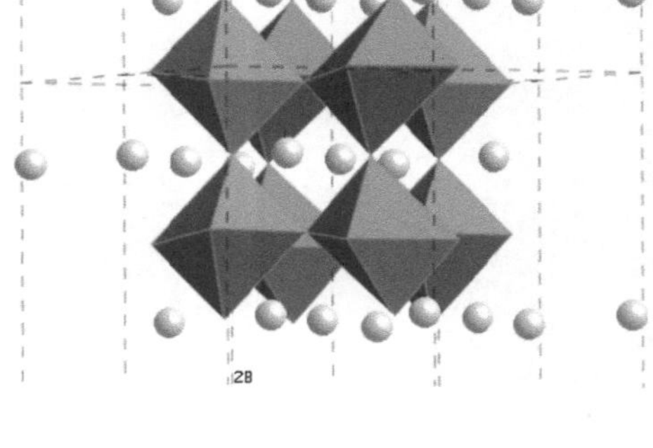

Fig. 1.2 Orthorhombic $GdFeO_3$ structure. Iron ions are octahedrally coordinated by the oxygen ions

La_2CuO_4 and La_2NiO_4 with the K_2NiO_4 structure are quasi-two dimensional compared to $LaCuO_3$ and $LaNiO_3$, which are three-dimensional perovskites [5]. Because of their varied features and properties, it has not been possible to establish a straightforward theoretical model to cover all TMO. However, there have been many convenient approaches to understand their electronic structures and physical properties. It is well established that the TMOs are strongly correlated electron system. In other words, the crystal structure, electrical, and magnetic properties are intimately related to each other. Hence, it is appropriate to have a brief overview of the synthesis, structure, properties, and their interrelation mechanism in these perovskites.

1.3 Syntheses of Few Perovskites

Polycrystalline perovskite manganites of the respective compositions could be prepared by the conventional solid-state synthesis, which is also considered to be easiest among all synthesis processes [5]. The method depends on the interdiffusion between the rare earth and transition metal oxide powders, and therefore it is necessary to use fine powders. The reaction also depends upon a few parameters, among them the most important being the diffusion length. In order to favor diffusion and obtain single-phase compounds, the reaction sintering temperature must be high so that the diffusion length 'l' exceeds the particle size, which is given by $l = \sqrt{(2kt)}$, where k = diffusion rate constant and t = sintering time. These parameters depend on the respective precursor materials [5]. Stoichiometric mixture of rare-earth oxide with respective alkaline earth carbonate and transition metal oxide thoroughly mixed in an agate mortar (sometimes with ethanol) and milled for few hours. After the powder mixed homogenously and dried, the mixture transferred to an alumina/platinum crucible and preheated to 1173–1373 K for 12–24 h with repeated intermediate grindings. The preheated powder ground thoroughly and palletized under a hydraulic pressure in a steel dye. The pellets/bars were then transferred into platinum boat and finally sintered at higher temperatures (1473–1673 K for 24–72 h) at different atmosphere, which depends upon the precursor materials as discussed below.

Polycrystalline perovskite cobaltites could be prepared by the conventional solid-state reaction method. Stoichiometric mixtures of the respective rare-earth oxides, $CaCO_3$, and Co_3O_4 were weighed in desired proportions and milled for few hours with propanol. After the mixed powders were dried, they were calcined in air at 1223 K followed by heating at 1273 and 1373 K for 12 h each in air. The powders thus obtained were pelletized and the pellets were sintered at 1473 K for 12 h in air. To improve the oxygen stoichiometry, the samples were annealed in an oxygen atmosphere at a lower temperature ($\leq$1173 K) and also sometimes in high pressures (80–130 bars). The oxygen stoichiometry of the perovskite samples could be obtained by iodometric titrations with their experimental limitations [5].

1.3.1 Single-Crystalline Perovskite Cobaltites

1.3.1.1 Floating Zone Melting Technique

The polycrystalline perovskite samples, discussed in the previous section, are easier to prepare as compared to single-crystal materials. In spite of that researchers are interested in preparation of high-quality single crystals which eliminate the impurities and defects in the material. There are several methods to grow single crystals: the Czochralski method, flux method, skull melting, and Bridgman and Stockbarger methods. The most popular technique for perovskite manganites and cobaltites' single crystal, however, is the floating zone melting technique. The advantage of this technique is that it does not require any container; therefore, the contamination from the container wall is completely avoided. Moreover, the uniform distribution of chemical constituents can be obtained by eliminating heterogeneous nucleation.

Figure 1.3 shows the schematic diagram of the floating zone image furnace. The furnace used in the present study was SC-M35HD, Nichiden Machinery Ltd., Japan. It consists of a pair of halogen lamps to generate infrared radiation, a pair of ellipsoidal mirror to focus the radiation onto the sample, a quartz tube enclosing the floating region for maintaining the required atmosphere, and two pulleys that can move independently or in a synchronized manner to control the growth rate. The growth rate can be varied from 0.1 to 99 mm/h. The maximum lamp temperature can be achieved up to 2400 K and the corresponding maximum pressure inside the

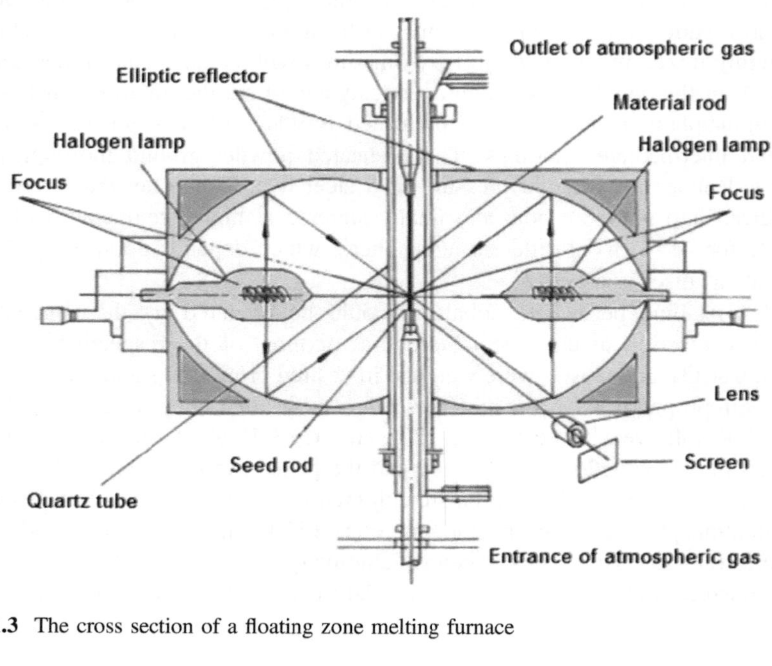

Fig. 1.3 The cross section of a floating zone melting furnace

quartz tube is of six atmospheric pressure. Infrared radiation coming from a pair of halogen lamps is focused onto the polycrystalline rods using a pair of gold-coated concave ellipsoidal mirrors.

Single crystals of perovskite cobaltites were grown by the floating zone furnace. The polycrystalline rods (feed and seed) were prepared by conventional solid-state reaction method, starting with stoichiometric mixtures of the rare-earth oxides, with $CaCO_3$ and Co_3O_4. The monophasic polycrystalline powders were hydrostatically pressed and sintered at 1473 K for 24 h in air to obtain feed and seed rods with a diameter of 3–4 mm and a length of 90–100 mm. Single crystals were grown under an oxygen flow of 2–4 L/min at a growth rate of 3–7 mm/h. A small part of the crystals was cut off and ground to fine powder on which an X-ray diffraction (XRD) was measured.

At the focal point the rods melt. The melts between the two connecting feed and seed rods are held by surface tension force without any other support and hence named floating zone melting technique. The stability of the molten zone can be controlled by changing the temperature, pressure, atmosphere, rod rotation speed, and molten zone pulling rate. Figure 1.4 shows the schematic presentation of different stages of the crystal growth using the floating zone furnace.

Figure 1.5 shows the image of a crystal grown by the floating zone melting technique, in an oxygen atmosphere with a growth rate of 7 mm/h. The crystal diameter is generally 4 mm and length is around 25 mm.

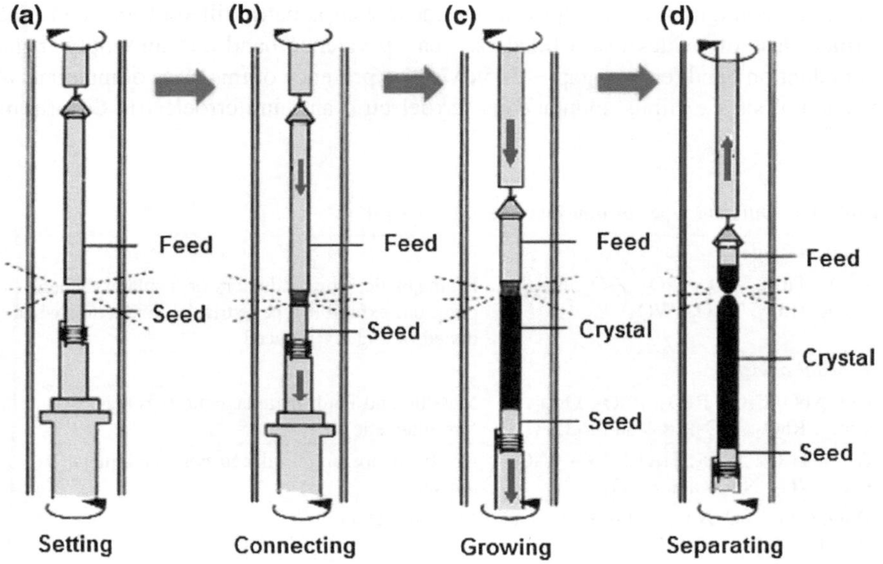

Fig. 1.4 Crystal growths inside the furnace

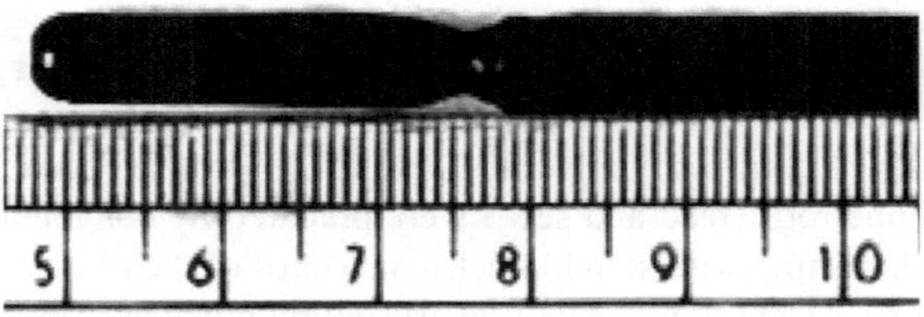

Fig. 1.5 Image of single-crystalline perovskite cobaltites $La_{0.7}Ca_{0.3}CoO_3$

1.4 Significant Properties of Perovskites

Perovskite oxides constitute an enormous group of solids exhibiting a wide variety of magnetic and electron transport properties [5]. These properties depend on the nature of the transition metal oxides present in the system. It is convenient to discuss the oxides of transition metals and non-transition metals separately. The examples of non-transition metal oxides are Na_2O, MgO, Al_2O_3, and SiO_2. Their electronic structure consists of a filled valence band (derived mainly from O:2*p*) and an empty conduction band (derived from the outer shells of metal atoms) separated by a large energy gap ($\sim$10 eV). They are therefore diamagnetic insulators under ordinary conditions [5]. Another type of perovskites is where two classes of transition metal ions are present in their structure: those in which the metal ion has d^0 electronic configuration and those where the *d* shell is partly filled (Table 1.1). The former class of oxides has a filled oxygen 2*p* valence band and an empty metal *d* conduction band (energy gap $\sim$3–5 eV). The presence of this class of materials at octahedral sites exhibits spontaneous ferroelectric and antiferroelectric distortions

Table 1.1 Different types of transition metal oxides [5]

d^0 metal oxides	
Sc_2O_3, TiO_2, V_2O_5, CrO_3, ZrO_2, Nb_2O_5, MoO_3, HfO_2, Ta_2O_5, WO_3	Diamagnetic semiconductors or insulators when pure, but exhibit *n*-type extrinsic conduction when doped or slightly reduced
d^n metal oxides	
TiO, NbO, CrO_2, ReO_2, RuO_2, OsO_2, MoO_2, RhO_2, WO_2, IrO_2 and ReO_3	Metallic and Pauli paramagnetic (CrO_2 is ferromagnetic)
Ti_2O_3, Ti_3O_5, Ti_4O_7, Ti_5O_9, V_2O_3, V_3O_5, V_4O_7, VO_2, NbO_2 and Fe_3O_4	Exhibit temperature-induced nonmetal–metal transition
MnO, FeO, CoO, NiO, Cr_2O_3, Fe_2O_3 and Mn_3O_4	Mott insulators
f^n metal oxides	
PrO_2, Ln_2O_3 (Ln = rare earth), $Pr_n O_{2n-2}$, $Tb_n O_{2n-2}$ and EuO	Insulators or hopping semiconductors. Paramagnetism characteristic of f^n configuration. EuO shows nonmetal–metal transition

useful for multiferroics. At high temperature these materials show intrinsic electron conduction, which may be due to oxygen loss or insertion of electropositive metal atoms into these oxides [5]. Transition metal oxides with partly filled *d* electron can be metallic or semiconducting. Some of them show temperature-induced nonmetal to metal transitions (Table 1.1). Magnetic properties also vary over a wide range from Curie–Weiss paramagnetism through spontaneous magnetism to Pauli paramagnetism. Metal oxides with d^n electronic configuration exhibit metallic properties when the overlap between orbitals of the valence shells of constituent atoms is large. Two kinds of metallic behavior can be distinguished: one due to strong cation–cation interaction and other due to strong cation–anion–cation interaction [5]. In Table 1.1, we have also included typical rare-earth oxides containing localized $4f^n$ electrons.

A number of isostructural TMO families exhibit perovskite, spinel, corundum, rutile, and other structures, which also show systematic changes in electron transport properties. We shall discuss the properties of perovskite oxides in some detail. The perovskite structure is ideally suited for the study of 180° cation–anion–cation interaction of octahedral site cations as shown in Fig. 1.6. The possibility of cation–cation interaction is remote because of the large interaction distance along the face diagonal. The variety in the properties of perovskites is illustrated by the following examples: $BaTiO_3$ is ferroelectric, $SrRuO_3$ is ferromagnetic, $LaFeO_3$ is weakly ferromagnetic, and $BaPb_{1-x}Bi_xO_3$ is superconducting, while $LaCoO_3$ shows a nonmetal–metal transition. Several perovskite oxides exhibit metallic conductivity; typical examples are ReO_3, $LaTiO_3$, and $LaNiO_3$. Metallic conductivity in perovskite oxides is entirely due to strong cation–anion–cation interaction.

Fig. 1.6 Perovskite structure, showing the possibility of cation–anion–cation interaction along the cube edge

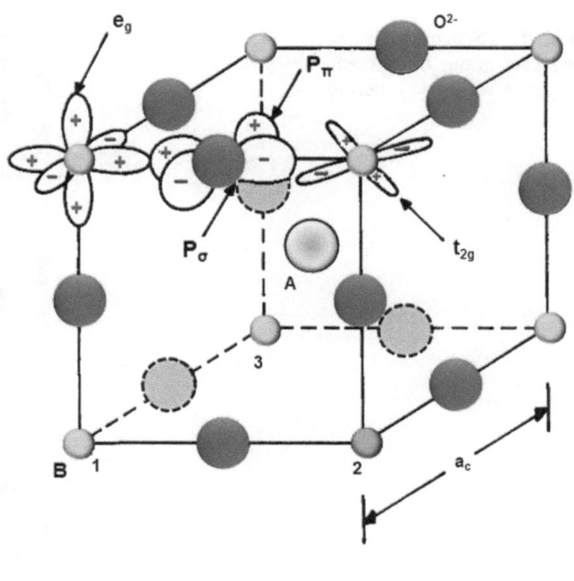

1.4.1 Magnetic and Electronic Properties

1.4.1.1 Magnetic Properties

The perovskite manganites such as $LaMnO_3$, $PrMnO_3$, and $NdMnO_3$ are insulator at all temperatures and undergo an antiferromagnetic (AFM) transition at low temperatures. The AFM ordering is of A-type, where ferromagnetically aligned *ab*-layers are coupled antiferromagnetically along *c*-axis. The different types of magnetic ordering are discussed below. The insulating nature of the parent compounds as well as the anisotropic magnetic interaction is related to their structure, in particular the JT distortion around Mn^{3+} ions. The A-site cation, as mentioned earlier, affects the JT distortion, in particular the Q_2 mode distortion which gives rise to long and short Mn–O distances, leading to the anisotropic exchange interaction. Removal of the JT distortion reduces the AFM interaction and this in turn destroys the AFM ordering.

Wollan and Koehler [10] studied the magnetic and crystallographic lattices in the series of manganates with the compositions, $Ln_{1-x}A_xMnO_3$, as a function of Mn^{4+} ion. The data obtained from neutron diffraction study of these manganates are in good agreement with Goodenough's predictions for different kinds of magnetic ordering.

The mixed-valence manganites (Mn^{3+} and Mn^{4+}) can exhibit magnetic ordering, charge ordering, and orbital ordering. Some of the simplest magnetic orderings for B-site cation are shown in Fig. 1.7, which represent one FM and four AFM (A-, C-, G-, and CE-type) ordering. In $Ln_{1-x}A_xMnO_3$ series, the magnetic lattice for $x = 1$ composition (i.e., $CaMnO_3$) corresponds to the G-type, whereas the $x = 0$ composition (i.e., $LaMnO_3$) has A-type AFM ordering. The other types are found across the

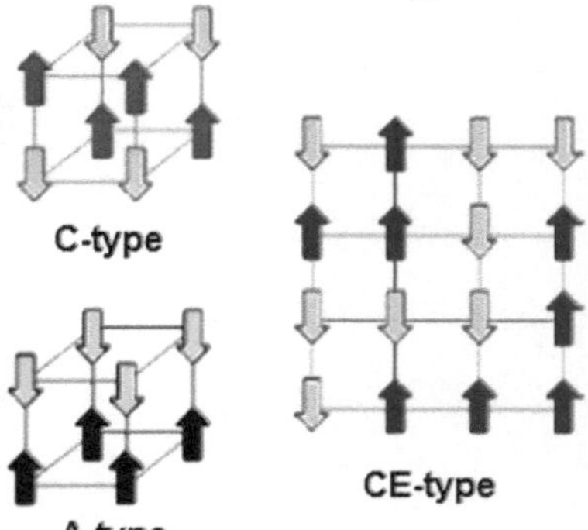

Fig. 1.7 Schematic representations of different types of magnetic ordering

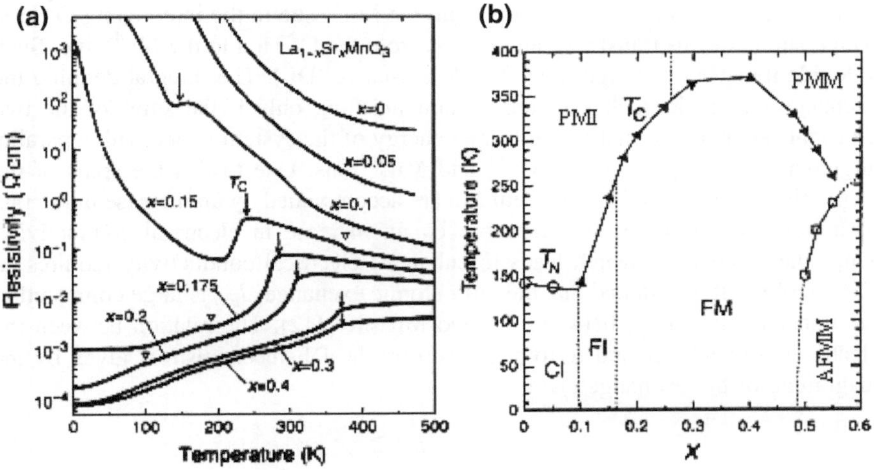

Fig. 1.8 **a** Temperature-dependent electrical resistivity and **b** magnetic phase diagram for $La_{1-x}Sr_xMnO_3$ perovskites (adapted from Ref. [11])

series as x varied, corresponding to the AFM ordering of Mn moments, which are C or CE-type. In C-type AFM, Mn ions order in alternate [111] planes, whereas in the CE-type they order in alternate [110] planes. When these perovskite manganites are hole-doped as in $Ln_{1-x}A_xMnO_3$ (Ln = La, Pr or Nd and A = Ba, Sr or Ca) the proportion of Mn^{4+} increases. The material becomes FM with a well-defined T_C at a finite value of x and also metallic below T_C. The FM and I–M transitions in the $La_{1-x}Sr_xMnO_3$ compositions are well defined [11] as shown in Fig. 1.8.

Although the compositions with $x > 0.5$ are essentially AFM, the FM clusters would also be present in the AFM phase. The material is generally ferromagnetic metallic (FMM) below T_C when $0.2 < x < 0.5$, and become paramagnetic insulator (PMI) when T > T_C. Thus, I–M transition occurring around T_C (Fig. 1.8) is well understood on the basis of the Zener double-exchange mechanism, which is discussed in the following section.

1.4.1.2 Zener Double-Exchange (DE)

The simultaneous observation of ferromagnetism and itinerant electron behavior (metallicity) in perovskite manganites is explained by Zener double-exchange model [12]. This involves the hopping of an electron from Mn^{3+} ($3d^4$, $t_{2g}^3e_g^1$; $S = 2$) to Mn^{4+} ($3d^3$, t_{2g}^3; $S = 3/2$) via oxygen ion where the Mn^{3+} and Mn^{4+} ions exchange takes place, i.e.,

$$Mn^{3+}-O^{2-}-Mn^{4+} \rightarrow Mn^{4+}-O^{2-}-Mn^{3+}$$

The transfer of an electron occurs from the Mn^{3+} site to the intervening O^{2-} ion with a simultaneous transfer of an electron from the O^{2-} ion to the Mn^{4+} site. Such a double transfer is referred to as double exchange (DE). The integral defining the exchange energy in such a system is nonvanishing only if the spins of the two *d*-orbitals are parallel, that is, the lowest energy of the system is one with a parallel alignment of the spins on the Mn^{3+} and Mn^{4+} ions. Due to this the spins of the incomplete *d*-orbitals of the adjacent Mn are accompanied by an increase in the rate of hopping of electrons and therefore by an increase in electrical conductivity. Thus, the mechanism which leads to enhanced electrical conductivity requires an FM coupling. It is assumed that the intra-atomic exchange, J_{ex}, is large compared to the transfer integral, t_{ij}, between the two Mn sites [12]. The relation between the electrical conductivity and ferromagnetism by the DE mechanism is given by the magnitude of the exchange energy, U_{ex}, as

$$U_{ex} = h\nu/2$$

where ν is the frequency of oscillation of the electron between two Mn sites and *h* is the Planck's constant. The diffusion coefficient for Mn^{4+} is related to exchange energy by

$$D = \alpha^2 U_{ex}/h$$

where α is the lattice parameter. Making use of the Einstein equation relating conductivity, σ, and D as

$$\sigma = ne^2 D/kT$$

where n is the number of Mn^{4+} ions per unit volume, one obtains

$$\sigma = xe^2 U_{ex}/\alpha hkT$$

Here x is the fraction of Mn^{4+} ions in $Ln_{1-x}A_xMnO_3$. Since in the FM transition T_C is related to the exchange energy by the approximate relation, $U_{ex} \approx kT_C$, one can write

$$\sigma \approx (xe^2/\alpha h)(T_C/T)$$

This equation relates the electrical conductivity to ferromagnetic T_C and the fraction of Mn^{4+} ions. We would therefore expect the IM transition in the manganites to occur at T_C. DE is strongly affected by structural parameters such as ⟨Mn–O–Mn⟩ bond angle or the Mn–Mn transfer integral (t_{ij}) [12].

The t_{2g} electrons of the Mn^{3+} ion are localized on the Mn site giving rise to a local spin of 3/2, but the e_g state, which is hybridized with the oxygen 2*p* state, can be localized or itinerant and only those electrons which have their spins parallelly aligned give rise to conductivity in the hopping process. There are strong Hund's rule interactions between the e_g and the t_{2g} electrons. Goodenough [13] pointed out that FM interaction is governed not only by the DE interaction, but also by the nature of the superexchange interactions. The magnetic exchange is strongly dependent on the structural distortion, as indeed shown by electronic structure calculations [12]. Recent investigations bring out the essential role of DE, along with the crucial role of the lattice and the electron–lattice interactions [12].

1.4.1.3 Superexchange Interaction

A pair of electrons of like spin, localized on degenerate orbitals, is lower in energy than a pair with opposite spins by an amount called the intra-atomic exchange energy. There are two classes of exchange interactions between spins on different atoms: (a) Direct exchange, which occurs between moments on atoms that are close enough to have significant overlap of their wave functions and decreases rapidly with increasing interatomic distance and (b) Indirect exchange, where the spin moments are coupled over relatively large distances. Indirect exchange manifests through an intermediary nonmagnetic ion (Superexchange) or through itinerant electrons (Ruderman–Kittel–Kasuya–Yoshida interaction; RKKY). Superexchange generally occurs in insulators, while RKKY coupling is important in metal [14].

Superexchange or AFM interaction between localized moments of ions in insulators that are too far apart to interact by direct exchange operates through the intermediacy of a nonmagnetic ion. Superexchange is able to occur when localized electron states as described by the formal valances are stabilized by an admixture of excited states involving electron transfer between the cation and the anion. The DE discussed above is different from superexchange, which also describes cation–anion–cation interactions. Different types of superexchange interaction are possible, depending on the structure of the oxide and the electronic configuration of the cations. Two important types are delocalized and correlation superexchange. Delocalized superexchange involves transfer of electron from one cation to another, as a result of cation–cation or cation–anion–cation interaction. Correlation superexchange is restricted to cation–anion–cation interaction. The Anderson–Goodenough–Kanamori rules[14] apply to superexchange interaction, according to which a 180° cation–anion–cation interaction in a d^3–O–d^3 system is AFM (Fig. 1.9a), but a d^3–O–d^4 interaction would be FM, whereas a 90° cation–anion–cation interaction between half-filled orbitals is FM, provided the orbitals are bonded orthogonally (Fig. 1.9b).

Superexchange involving σ bonds is stronger than those involving π bonds. In the 3d transition metal mono-oxides, the ordering temperature T_N increases in the order MnO < FeO < CoO, because the σ interaction increases in that order. For cations of the same electronic configuration, superexchange is stronger for the higher valency cation (e.g., Fe^{3+} > Mn^{2+}) [5].

Superexchange interaction would not, however, give rise to increased electrical conduction as in the DE mechanism. Zener's model has been extended or modified by several workers. In particular, de Gennes [15] has shown that the energy of the

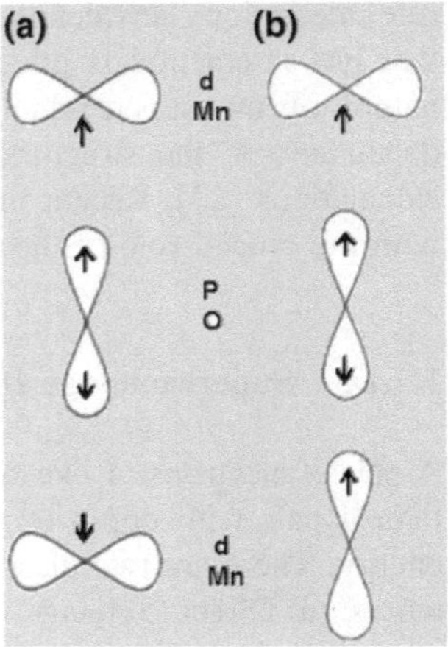

Fig. 1.9 A cation–anion–cation interaction in a Mn–O–Mn system

electrons get lowered if there is canting of the sublattices, giving rise to a canted spin AFM state, and a situation found in $La_{1-x}A_xMnO_3$ when x is small.

1.4.1.4 Electrical Properties

Electrical conductivity, thermal conductivity, Seebeck effect, and Hall effects are some of the common electron transport properties of perovskite oxides that characterize the nature of charge carriers. On the basis of electrical properties, perovskites may be classified into metals, semiconductors, and insulators, wherein charge carriers move in the band states [16]. The electron transport properties of perovskites provide useful criteria for distinguishing localized and itinerant electron in solids. In certain semiconductors and insulators, charge carriers are localized, and their motion involves a diffusive process. The semiconductor or insulator like transport behavior in perovskite oxides is characterized by three models defined by $\log \rho$ α $T^{-1/n}$ where $n = 1, 2,$ or 4; (i) Here $n = 1$ corresponds to a simple Arrhenius law, which describe the thermally activated behavior due to band gap or mobility edge, (ii) When $n = 2$, the hopping is referred to as Efros–Shklovskii-type hopping (ESH); and (iii) Here $n = 4$ corresponds to variable range hopping (VRH). The hopping dynamics is controlled by the collective excitation of the charge carriers [16]. To understand the transport mechanism in rare-earth perovskites the data was analyzed based on these three models.

The electron transport property of $Ln_{1-x}A_xMnO_3$ perovskites has one-to-one correspondence with the magnetic properties, exhibiting low resistivities in the FM region and semiconducting or insulating behavior in the CO and AFM regions. The manganites with $x < 0.5$ have a conduction band more filled than half filled, whereas those with $x > 0.5$ have conduction band less than half filled. The number of charge carriers in the system can be assumed to be equal to the Mn^{4+} ion introduced into the lattice for small doping in the A-site. The process involves the transfer of electrons from Mn^{3+} to Mn^{4+} by DE mechanism. Most of the perovskite manganites (parent compounds) are paramagnetic insulators at room temperature and exhibit an increase in electrical resistivity with the decrease in temperature. Compositions that are FM show insulating behavior above T_C, but with decreasing temperature (below T_C) the resistivity decreases as in the metals. This I–M transition is therefore associated with a peak in resistivity at a temperature T_{IM}. Generally, T_{IM} is somewhat lower than T_C and the sharpness of the transition in polycrystalline samples as well as in films often depends on the sample quality.

1.4.2 Colossal Magnetoresistance (CMR)

The change in electrical resistance of a material in response to an applied magnetic field is referred to as magnetoresistance (MR). In general, MR is defined by

$$MR = [\rho(H) - \rho(0)]/\rho(0) = [\Delta\rho/\rho(0)]$$

where ρ(H) and ρ(0) are the resistivities in the presence and absence of magnetic field (H), at a particular temperature. MR can be positive or negative depending on the materials. In magnetic material, MR is negative, because of the suppression of spin disorder by the magnetic field. Many solids exhibit small MR owing to the Lorenz force that a magnetic field exerts on moving electrons which is known from physics text book since 1950. Very large MR, referred to as giant magnetoresistance (GMR), was first reported by Baibich et al. [17] in layered Fe/Cr metallic multilayer. Large MR was observed in powder and single crystals of doped perovskite manganites of the type $Ln_{1-x}A_xMnO_3$ in mid 1980s. But the renewed interest in these perovskites started with the report of negative MR in $Nd_{0.5}Pb_{0.5}MnO_3$, [18] which showed MR of 50 % near T_C (184 K). This report was followed by studies on thin films by several other workers [19]. In perovskite $La_{0.67}Ca_{0.33}MnO_3$, MR was found to be extremely large (1,000-fold change in the resistance for a few tesla magnetic field) and hence termed as colossal magnetoresistance (CMR) by Jin et al. [19] The discovery of negative GMR in perovskite manganites has attracted wide attention. The magnitude of negative GMR in these materials can be very large, close to 100 %. Therefore, many researchers prefer to call it CMR, as distinct from GMR in layered or granular metallic materials [20]. In metallic multilayers or granular alloys, the mechanism involves spin-polarized transport. Also, in perovskite manganites spin-polarized transport is responsible for the large negative MR, but it is distinctly different from the metallic multilayers.

The application of a magnetic field ($\sim$few Tesla) leads to a significant decrease in the resistivity of perovskite manganites; the magnitude of decrease in resistivity (i.e., MR) is highest in the region of T_C or T_{IM}. A typical example of perovskite $La_{1-x}A_xMnO_3$ is shown in Fig. 1.10, where the magnetization, resistivity, and CMR are plotted as a function of temperature at various fields. The highest CMR effect is

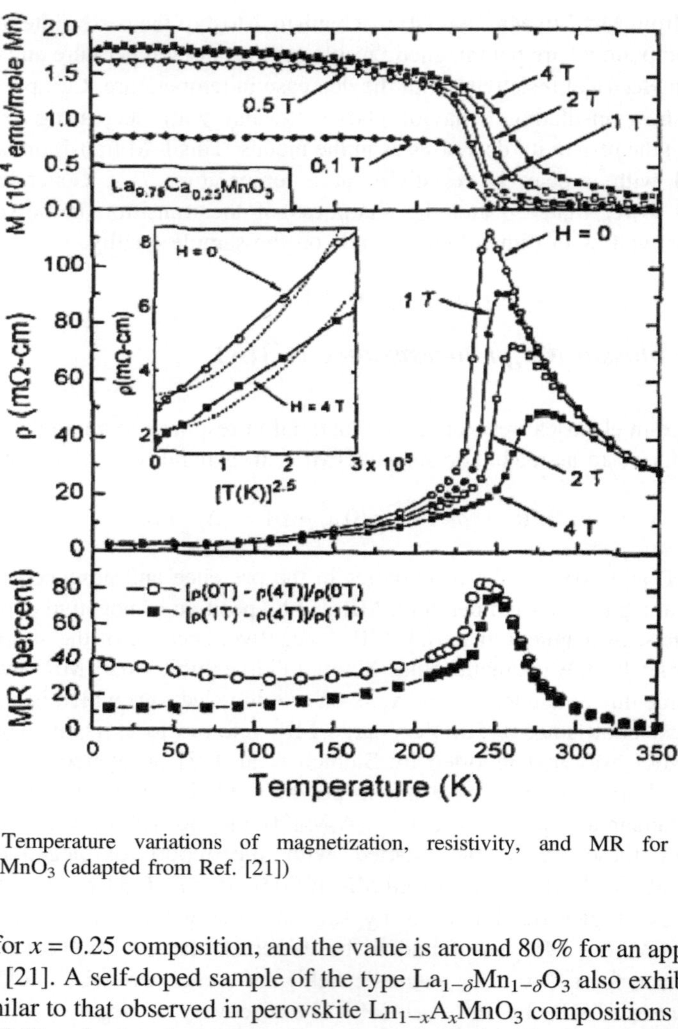

Fig. 1.10 Temperature variations of magnetization, resistivity, and MR for perovskite $La_{0.75}Ca_{0.25}MnO_3$ (adapted from Ref. [21])

observed for $x = 0.25$ composition, and the value is around 80 % for an applied field of 4 Tesla [21]. A self-doped sample of the type $La_{1-\delta}Mn_{1-\delta}O_3$ also exhibits CMR effects similar to that observed in perovskite $Ln_{1-x}A_xMnO_3$ compositions [22]. The effect of CMR and related properties have been studied in the Ruddlesden–Popper phases by Moritomo et al. [23] and Mahesh et al. [23], in $(SrO)(La_{1-x}Sr_xMnO_3)_n$ compound. Another class of compound is the ordered perovskite (Sr_2FeMoO_6), which exhibits tunneling magnetoresistance (TMR) at room temperature [24]. Other than the perovskite manganites, CMR is also found in $Ti_2Mn_2O_7$ (with pyrochlore structure) which has only Mn^{3+} ions [25].

In perovskite cobaltites $La_{1-x}A_xCoO_3$ (A = Ba, Sr, Ca and Pb), the studies of Briceno et al. in 1995 revealed significant MR [26]. Unlike the manganite perovskites, there are less studies on the MR effect of cobaltites and most of them are focused on $La_{1-x}Sr_xCoO_3$ perovskite only. The MR effects are very strongly connected to the magnetic states of the system. A very intriguing fact is that the metallic compositions of $La_{1-x}Sr_xCoO_3$ with $x > 0.2$ do not show sizable MR effect

[26]. For the metallic samples, the MR of $La_{0.85}Sr_{0.15}CoO_3$ single crystal near the metal–insulator transition exhibits a typical non-hysteretic negative MR (20 %) in the vicinity of T_C as reported by J Wu et al., and the MR increases with an increase in the cluster size [26]. It, however, does not occur below 35 K, where the cluster size decreases and a sharp increase in MR takes place. A hysteretic large negative MR is noticed (~68 %), which persists up to a high field. It was proposed that the spin-dependent transport between FM clusters gave rise to the hysteretic feature of the low-temperature MR. The MR of $Ln_{1-x}A_xCoO_3$ cobaltites is significantly influenced by the substitution at the cobalt site.

In perovskite cobaltites, when cobalt is replaced by a nonmagnetic ion Ga^{3+} in $La_{0.7}Sr_{0.3}Co_{1-x}Ga_xO_3$, a sudden increase in the MR effect on the insulating phase at low temperature is reported by Wang et al. [26] due to the suppression of spin disorder by application of magnetic field. The MR of $La_{0.5}Sr_{0.5}Co_{1-x}Ru_xO_3$ perovskite increases up to 40 % at 50 K for $x = 0.1$; however, there is no significant change in the MR for x > 0.1 as reported by Hsu et al. [26]. The alignment of the canted spin structure results in an improved MR by allowing the electrons to hop more easily in an applied magnetic field. It was also reported by Maignan et al. [26] that the MR enhanced up to ~60 % for $x = 0.10$ in $La_{0.8}Sr_{0.2}Co_{1-x}Mn_xO_3$. The result was explained by the decrease in ferromagnetism and the increase in resistivity by Mn doping, resulting from the progressive replacement of mobile holes on Co^{4+} by localized holes on Mn^{4+}.

The investigation of large MR in layered 112-type $LnBaCo_2O_{5.4}$ cobaltites (Ln = Eu and Gd) by Raveau's group [27] in 1997 has inspired many studies in order to understand its origin. The negative MR below T_N in the 112 cobaltite is coupled to the AFM order. The origin of this effect may be associated with the competition between FM and AFM that appears in these systems. There are merely few reports that focus on the magnetoresistance in these 112 cobaltites. Maignan et al. [27] observed 41 % MR in $LnBaCo_2O_{5.4}$ cobaltites and suggested to be associated with the complex magnetic behavior of the systems. A magnetic field that supports one kind of ordering in the systems with competing magnetic orders often results in a large MR. Although a similarity exists between the MR effect of cobaltites and that of manganites, nevertheless it is noteworthy that the cobaltites remain insulators even in the FM state, illustrating the irrelevance of either the double-exchange mechanism or the percolation through some metallic phase. There are reports of MR effect in $GdBaCo_2O_{5.5\pm\delta}$ being not even related to the spin valve effects, that is, from a tunneling between two spin-polarized metallic regions as reported by Taskin et al. [27]. An ample anisotropy has indeed been observed in $GdBaCo_2O_{5.5}$ in the isothermal magnetoresistance with regard to the field direction. With the application of a field perpendicular to the *c*-axis, a hysteretic negative MR is found to occur. The relative MR demonstrates a robust anisotropy of −93 and −22 % for the field perpendicular as well as parallel to the *c*-axis, respectively, as reported by Zhou et al. [27]. A magnetic field dependence of MR at different temperatures is observed near the ferro/antiferromagnetic phase boundary for the first member of the 112 layered cobaltite $LaBaCo_2O_{5.5}$. The highest MR value, about −5 %, is observed near the FM–AFM phase boundary as pointed out by Kundu et al. [27].

1.4.3 *Effect of Cation Size and Disorder on Properties*

In perovskite manganites various interesting physical phenomena are governed by the width of the e_g band, which is directly determined by the average radius of A-site cation $\langle r_A \rangle$ or the tolerance factor. This is because a distortion of the ⟨Mn–O–Mn⟩ bond angle affects the transfer interaction of the e_g conduction electrons (holes). In the manganites, the T_C increases with $\langle r_A \rangle$, whereas T_{CO} increases with the decrease in $\langle r_A \rangle$. The increase in $\langle r_A \rangle$ is equivalent to increase in the external hydrostatic pressure and is therefore accompanied by an increase in the ⟨Mn–O–Mn⟩ angle and the e_g bandwidth. The sensitivity of T_{CO} to $\langle r_A \rangle$ is studied by several workers [28] and is generally attributed to an increase in tilting of the MnO_6 octahedra as the $\langle r_A \rangle$ decreases.

Figure 1.11 shows a schematic phase diagram of the rare-earth perovskite manganites, which describe the FM metal, AFM insulator, spin and charge-ordered insulating state with the variation of $\langle r_A \rangle$ [29]. In region A, when $\langle r_A \rangle$ is large (e.g., $La_{1-x}Sr_xMnO_3$), only ferromagnetism and the associated IM transition occur (with no CO). With a slight decrease in $\langle r_A \rangle$ as in region B, the FM metallic state transforms to AFM charge-ordered state ($T_{CO} = T_N$) on cooling the system (e.g., $Nd_{0.5}Sr_{0.5}MnO_3$, $\langle r_A \rangle$ = 1.236 Å). When $\langle r_A \rangle$ is sufficiently small (region D) as exemplified by compounds $Pr_{0.7}Ca_{0.3}MnO_3$ with $\langle r_A \rangle$ of 1.17 Å, [30] no ferromagnetism is encountered and CO occurs in the paramagnetic state. Depending on

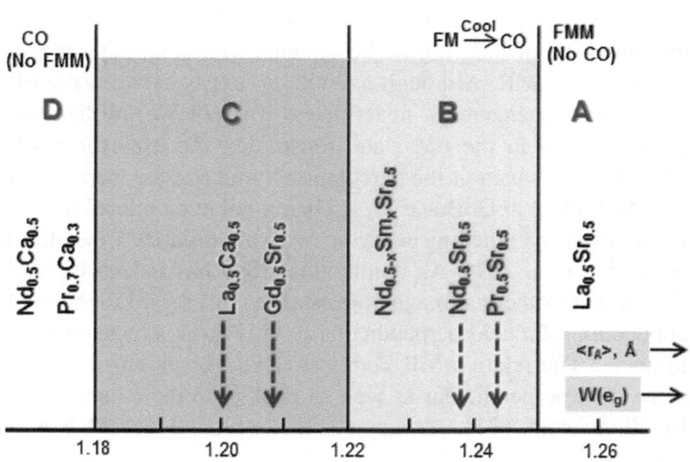

Fig. 1.11 Schematic diagram showing the prevalence of charge ordering and FM states in manganites depending on the $\langle r_A \rangle$ or the e_g bandwidth (adapted from Ref. [29])

the $\langle r_A \rangle$ value, the CO state can be melt to FM metallic state by the application of magnetic field. However, in $Y_{0.5}Ca_{0.5}MnO_3$ ($\langle r_A \rangle \sim 1.13$ Å), the charge-ordered state ($T_N < T_{CO}$) is robust and is not affected at very high magnetic field (>25 T) [28]. In region C, the perovskite manganites show rather complex behavior; by the variation of $\langle r_A \rangle$, one can bring manganites in the region of B and D to region C. Thus, in $La_{0.25}Nd_{0.25}Ca_{0.5}MnO_3$ with $\langle r_A \rangle = 1.19$ Å, on cooling the system a novel re-entrant FM transition occurs from a CO state [31]. There is coexistence of two phases in the temperature range of 150–220 K, around the CO–FM metallic transition. Thus, the formation of FM clusters in an AFM–CO matrix in the manganites gives rise to interesting magnetic properties like CMR, electronic phase separation, spin glass, etc. The magnetic and electrical properties of the manganites are consistent with the occurrence of electronic phase separation and glassy magnetic behavior corresponding to a critical average radius $\langle r_A^c \rangle$ of 1.18 Å [32].

The effect of cation size disorder on the various physical properties of perovskite manganites has been reported by several workers. The ferromagnetic T_C and insulator–metal transition T_{IM} increases with increase in $\langle r_A \rangle$. However, if there is considerable mismatch in the radius of the different A-site cations, then the T_C does not increase with increase in $\langle r_A \rangle$ as shown in Fig. 1.12. The size disorder effect arising from the mismatch of the A-site cation on the T_C or T_{IM} has been analyzed by the σ^2 parameter, where σ^2 is defined as $\sigma^2 = \sum x_i r_i^2 - \langle r_A \rangle^2$. Here x_i ($\sum x_i = 1$) is the fractional occupancy of the A-site ions, r_i is the corresponding ionic radii, and $\langle r_A \rangle$ ($\langle r_A \rangle = \sum x_i r_i$) is the weighted average radius calculated from r_i values [33]. The T_C decreases significantly with increase in variance σ^2, which is reported for both the perovskite manganites and cobaltites by several workers [34]. A similar study of the variation of T_{CO} with σ^2 in manganites for fixed $\langle r_A \rangle$ values of 1.17 and 1.24 Å has shown that T_{CO} is not very sensitive to size mismatch [34].

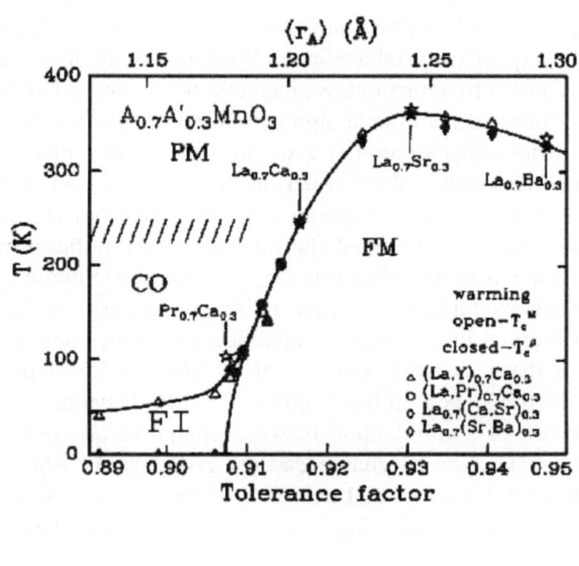

Fig. 1.12 Magnetic phase diagram of the $Ln_{0.7}A'_{0.3}MnO_3$ series of perovskite manganites (adapted from Ref. [34])

In the perovskites, those containing Ba in the A-site show an exceptional potential for the generation of specific properties, inducing interesting magnetic properties. This behavior originates from the large size of Ba, compared to other A-site cations, such as Ca, Sr, or Ln. Indeed, the simultaneous presence of Ba and of another cation in the A-site of the perovskites is susceptible to introduce large distortions of the $[BO_3]$ framework, due to the size difference between the A-site cations. This phenomenon could be overcome by the introduction of vacancies on the oxygen sub-lattice, requiring the B-site element to accommodate a coordination number smaller than six. As a result oxygen-deficient perovskites can be stabilized by simultaneous ordering of the cationic and anionic sites. The $LnBaM_2O_5$ (M = Mn, Co, Fe)-type-layered perovskites illustrate the impact of the structure and oxygen stoichiometry upon the physical properties of these perovskites [35]. The comparison of the magnetism of the order layered $LaBaMn_2O_6$ with that of the disordered perovskite $La_{0.5}Ba_{0.5}MnO_3$ shows a remarkable feature: The T_C is increased from 270 K for the disordered phase to 335 K for the ordered one. This view point is strongly supported by the magnetotransport properties of the isotypic perovskite cobaltites [35].

1.4.4 Electronic Phase Separation in Perovskites

The coexistence of more than one phase at a particular condition is referred to as phase separation; a new phenomenon recently found to occur in certain transition metal oxides [3, 36]. The phase separation gives rise to electronic inhomogeneity and is associated with a diverse variety of electronic and magnetic properties. This is not expected for nominal monophasic compound. This is due to nonuniformity in impurity distribution, as it can have an electronic origin or could also arise from the presence of magnetic impurities. Intrinsic inhomogeneities are present even in the best quality crystal available. Such electronic phase separation can be controlled or changed by temperature, magnetic fields, and other external factors. In these types of phase separation, a high carrier density favors FM ordering and/or metallicity. If carrier concentration is low, the FM metallic phase can occur in one part of the crystal keeping the rest part as an insulating and AFM phase.

Impurity phase separation is different from electronic phase separation (EPS), and there is no mutual charging of phases in the former. The diffusion of impurity atom has to be sufficiently large to give rise to phase separation in such system. One such example is the case of oxygen-excess La_2CuO_4 [36]. The EPS has been observed in magnetic semiconductors such as heavily doped EuSe and EuTe [36]. In these systems, the crystal is AFM at low temperatures and the conducting electrons occur in the droplets. The EPS is entirely reversible, and in general is the result of a competition between charge localization and delocalization. The large concentration of charge carriers gives rise to FM metallic state in a part of the crystal causes mutual existence of two phases. At relatively small carrier concentration, the conducting FM regions are separated and form droplets (Fig. 1.13a, b)

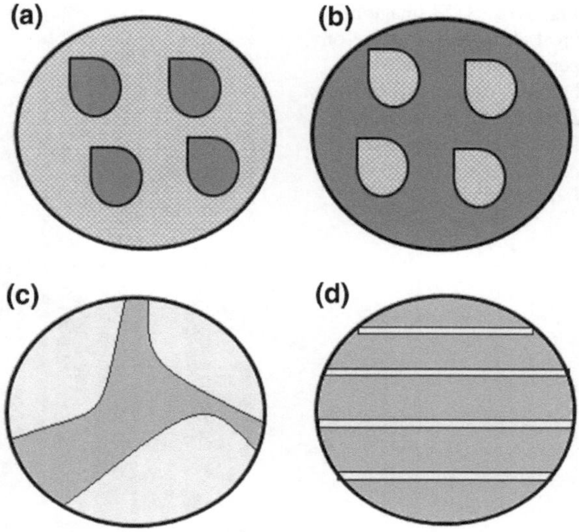

Fig. 1.13 Schematic representation of microscopic electronic phase separation between ferromagnetic metallic (*blue*) and non-ferromagnetic insulating state (*green/yellow*)

or random/stripes (Fig. 1.13c, d). With increasing carrier concentration, the volume of the FM phase increases rendering the droplets to coalesce and gives rise to a situation as shown in Fig. 1.13. An interesting phenomenon of EPS is that it covers a wide range of length scales anywhere between one to a few hundred of nanometer. The phase separation on a larger length scale is not possible because of strong Coulomb energy. In the presence of Coulomb interaction, the microscopically charge-ordered state is stabilized giving rise to clusters of one phase embedded in another. The size of the clusters depends on the competition between DE and Coulomb forces. In phase separation, the phases of different charge densities are generally expected to give rise to nanometer (1–200 nm)-scale clusters. This is related to the larger phase-separated domains which would break up into small pieces because of Coulomb interactions. The JT distortion associated with the Mn^{3+} ions and CO of Mn^{3+} and Mn^{4+} ions compete with DE interaction and promote the insulating AFM behavior [36].

The techniques used to probe the phase separation in different length scales (Fig. 1.13) are scanning probe microscopy, atomic force microscopy, neutron diffraction, NMR, and Mossbauer spectroscopies. Figure 1.14 shows the presence of electronic inhomogeneities in the perovskite $La_{1-x}Ca_xMnO_3$ manganites observed by scanning tunneling microscope (STM) [37]. The dark-field electron microscopy images of $(La_{1-y}Pr_y)_{1-x}Ca_xMnO_3$ show a clear competition between CO-AFM versus FM phase, which can be tuned by varying the relative amount of La and Pr [32]. Renner et al. [38] have reported the evidence of phase separation even at room temperature. High-resolution X-ray and neutron diffraction investigations have shown that in $Nd_{0.5}Sr_{0.5}MnO_3$ manganites three macroscopically different phases coexist at low temperatures [39].

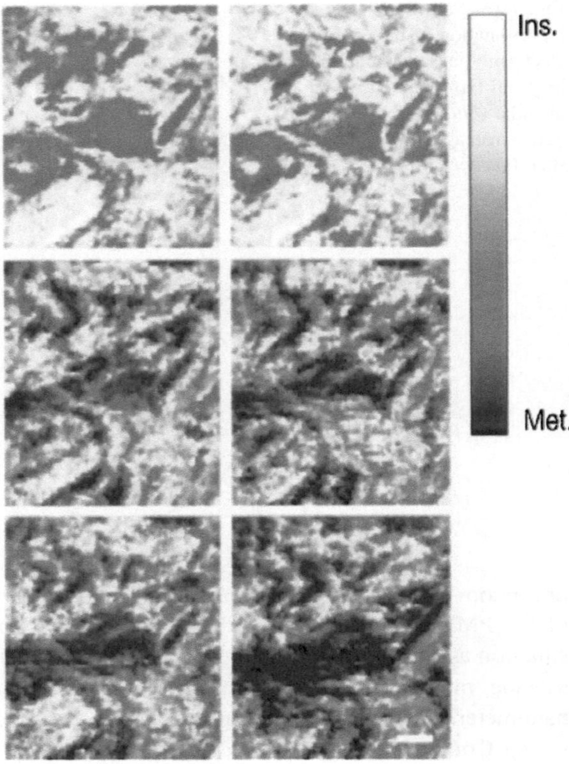

Fig. 1.14 STM image of the local electronic structure of perovskite $La_{1-x}Ca_xMnO_3$ below T_C (scale bar: 100 nm). *Light colors* represent the insulating, whereas *dark colors* are metallic region (adapted from Ref. [37])

The existence of EPS in perovskite cobaltite at low temperature has been reported by several researchers for last few decades [40]. It is also worth mentioning that the hole-doped $Ln_{1-x}A_xCoO_3$ perovskite cobaltites have been reported to exhibit electronic phase separation in 1967 by Raccah et al. [40] and there was no technological advantage for further pursuit. The EPS is evidenced in these perovskite cobaltites due to the competition between Co^{3+}–O–Co^{3+} and Co^{4+}–O–Co^{4+} AFM interactions and the Co^{3+}–O–Co^{4+} FM interactions, where hole-rich FM clusters are embedded in a hole-poor insulating matrix. At low x values, the cobaltites are dominated by AFM interactions between Co^{3+} cations. With the increase in x, Co^{4+} cations appear, and EPS takes place, where the FM hole-rich clusters inhabit within the AFM or nonmagnetic matrix. The existence of inhomogenous distribution of La-rich (hole-poor) and Sr-rich (hole-rich) regions with sizes 8–40 nm has been evidenced through high-resolution electron microscopy studies by Caciuffo et al. Additionally, the ^{59}Co nuclear magnetic resonance (NMR), neutron diffraction, and small-angle neutron scattering provide strong evidence for EPS in perovskite cobaltites [40].

The magnetism in $Pr_{1-x}Ca_xCoO_3$ seems to be inhomogeneous for $x < 0.3$ and develops into homogeneous type with increasing x as reported by Tsubouchi

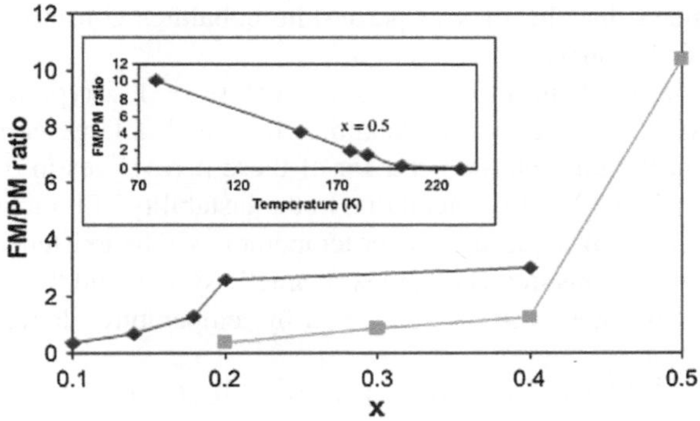

Fig. 1.15 Variation of the ratio of the ferromagnetic to paramagnetic species with composition $La_{1-x}Sr_xCoO_3$: squares, Mössbauer data at 78 K from Bhide et al. [40] diamonds, NMR data at 1.9 K from Kuhns et al. [40]. The inset shows the temperature variation of the FM/PM ratio of $La_{0.5}Sr_{0.5}CoO_3$ taken from the Mössbauer data (Adapted from Kundu et al. [40])

et al. [40]. Imada et al. [40] pointed out that, in addition to the well-known spin state transitions, EPS occurs arising from the comparable sizes of the Hund's rule exchange energy and crystal field splitting. Investigations on heat capacity and SANS in combination with statistical simulation in $La_{1-x}Sr_xCoO_3$ revealed that the phase separation is driven solely by inevitable local compositional fluctuations at nanoscopic length scales, rather than being electronically driven. He et al. [40] asserted that more complex EPS models are not required to understand the observed phenomena in $La_{1-x}Sr_xCoO_3$ perovskite cobaltite. Evidence of an inhomogeneous magnetic ground state in single crystals of $La_{1-x}A_xCoO_3$ (A = Ca^{2+}, Sr^{2+} and Ba^{2+}) due to the competing FM and AFM interactions has been recently provided by Yu et al. [40].

Figure 1.15 shows a variation in the FM to paramagnetic (FM/PM) ratio with the composition for $La_{1-x}Sr_xCoO_3$ cobaltites. The FM/PM ratio increases with x. However, even at low temperatures, the PM feature continues to exist as well for $x = 0.5$ composition. The PM phase continues to exist much below $T_C = 220$ K, nevertheless the FM/PM ratio increases with a decrease in temperature. Additionally, the relative proportion of the FM to the PM species is sensitive to the Ln size as reported by Kundu et al. [40]. A double FM transition in $Pr_{0.5}Sr_{0.5}CoO_3$ was also proposed by Mahendiran et al. [40] to be associated with the electronic or structural phase separation. The huge coercive fields and large thermomagnetic irreversibilities of the $La_{0.7}Ba_{0.3}CoO_3$ compound have been interpreted by Ganguly et al. [40] on the basis of a possible coexistence of different magnetic phases. Studies have been conducted on the influence of $\langle r_A \rangle$ and size disorder parameter, σ^2, upon phase separation for a large number of cobaltites by Kundu et al. [40]. It has been demonstrated that the EPS tendency increases with the decrease in $\langle r_A \rangle$, whereas it decreases with decreasing σ^2. A similar observation was also reported in different perovskite cobaltites and manganites by Rao et al. [3]. Thus, most of these studies recommend that the EPS in perovskite cobaltites consists of FM clusters inside the non-FM matrix.

The EPS is amongst the most interesting feature of 112-type-layered oxygen-deficient $LnBaCo_2O_{5.5\pm\delta}$ perovskite cobaltites for which a large number of studies take into account the multiphasic behavior of these perovskites in order to explain the experimental facts [27]. In general due to the instability of homogeneous carrier distribution, the EPS takes place at lower temperatures. The existence of competing FM and AFM interactions in this complex magnetic system and the delicate balance between these two states is strongly affected by temperature, doping, or magnetic field.

Taskin et al. [27] have proposed that the $GdBaCo_2O_{5.5\pm\delta}$ cobaltite exhibits interesting nanoscopic phase separation into two insulating phases in the electron-doped regime, while an insulating and a metallic phase is marked in the hole-doped region. The coexistence of FM and AFM phases between 210 and 150 K in $GdBaCo_2O_{5.5\pm\delta}$ has been demonstrated by García-Fernández et al. [27] through the diffraction experiments. Similar results have been reported for both $NdBaCo_2O_{5.5}$ and $TbBaCo_2O_{5.5}$ cobaltites [27]. The thermoelectric and transport properties of $LaBaCo_2O_{5.5}$ cobaltite have been explained in terms of the coexisting FM and AFM phases by Kundu et al. [27]. The EPS and charge ordering have been simultaneously observed in $YBaCo_2O_5$. The EPS is triggered by the relaxation of the lattice distortion due to the charge ordering as reported by Akahoshi et al. [27]. A phase separation scenario was suggested in differently ordered perovskites (Pr, Sm, and Eu phases) to explain the magnetic data by Seikh et al. [27], where the FM domains are embedded within the AFM matrix. An elaborate investigation of EPS in perovskite cobaltites $EuBaCo_{1.92}M_{0.08}O_{5.5\pm\delta}$ (M = Zn, Cu, and Ni) has been performed by Raveau et al. [27]. It has been reported that at low-temperature FM and super-paramagnetic droplets are embedded in an AFM matrix. Additionally, Seikh et al. [27] have proposed that, the Ca-doped $EuBaCo_2O_{5.50\pm\delta}$ cobaltites exhibit an EPS scenario involving the canted AFM domains in the high-temperature paramagnetic phase. Recently, in Ca-doped $YBaCo_2O_{5.5}$ perovskite cobaltites, the authors have suggested about the existence of EPS involving Co^{3+} FM clusters and Co^{2+}AFM clusters [27]. Moreover, Sarkar et al. [27] have suggested that the low temperature state is not in thermal equilibrium. The glassy state is formed due to the presence of EPS with the assistance of an external field, which makes it distinctly different from the spin glass state.

In order to understand the magnetic and electron transport properties, CMR effect, magnetic glassy behavior, etc. of magnetic perovskites, EPS is getting accepted as the phenomenon of importance [3]. The modern technologies such as scanning electron nanodiffraction, [36] atomic-resolution STM, MFM, and electron holography have been developed, which directly identify the EPS phenomenon in low-dimensional perovskite manganite nanostructures [38]. Along with the development of nanotechnology, the EPS phenomenon in perovskite-based CMR nanoparticles has also received great attention. In recent years, the evolution of the EPS with magnetic field in perovskite nanoparticles has been reported by several groups, which has a significant impact on the perovskite-based nanoelectronics [38].

1.4.5 Spin Glass Behavior in Perovskites

The spin glass represents a non-equilibrium state which is a rather complex kind of condensed state in solid-state physics. In 1970s the spin glasses were all important, and enormous effort was made to measure and explain the unique freezing (an unconventional spin glass transition) and the low-temperature glassy behavior [41]. In the following years experimental and theoretical studies have revealed some strong support of this new magnetic phenomenon, which are associated with the frustration and disorderliness of the magnetic system [42]. The nature of this new kind of material raises many fundamental questions and thus its complete theoretical description is still under discussion.

In brief, the spin glass material can be described as, a magnetic system in which the interactions between the magnetic moments are "in conflict" with each other, due to some frozen-in structural disorder. Therefore, there is an absence of conventional long-range ordering (FM or AFM type) in these systems. Thus, the spin glasses consist of an ensemble of disorder spins, and represent a model system for the statistical mechanics of a system with quenched randomness. Nevertheless, these systems exhibit a "freezing transition" to a state with a new kind of "order," where the spins are aligned in random directions. The actual spin ordering in the spin glass is a problem belonging to the physics of structurally disorder materials, and does not arise in more conventional regular systems. Because of the spin disorder at low temperature, the spins are subject to different types of interactions like FM (positive) or AFM (negative). In this situation a particular spin will receive conflicting information on the way of ordering from its nearest neighbors and therefore it will not be possible for the system to arrange in a certain spin configuration to minimize its energy. This phenomenon is commonly known as frustration, [43] which is shown schematically in Fig. 1.16.

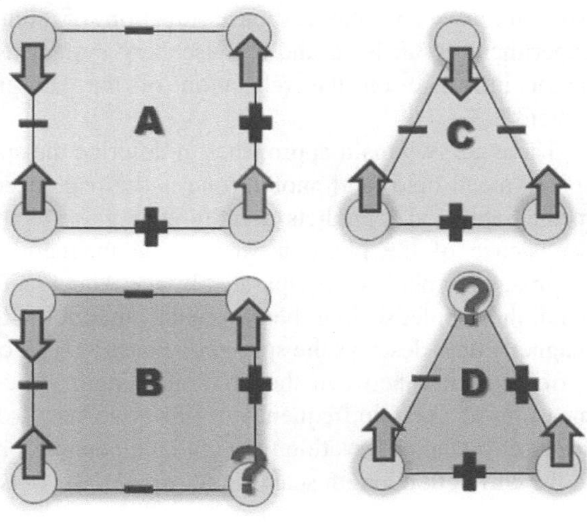

Fig. 1.16 Schematic representations of spin glass behavior. The positive and negative signs represent FM and AFM interactions

Figure 1.16A represents a square lattice without frustration, since all positive and negative interactions are satisfied. The spin on the upper left couples antiferromagnetically with the spins on the upper right and lower left, while the spin on the lower right couples ferromagnetically with them. In Fig. 1.16B, the frustration appears, since there is no even number of positive and negative spins. Therefore, the frustration originates from the disorder of the interactions. Figure 1.16C represents different spin arrangements in the triangular lattices. In this case, there is no magnetic disorder, since all sites are occupied and there is no frustration, but it appears in Fig. 1.16D as in square lattice.

In order to perform experiments on SG, first of all it is necessary to make sure that the given system does not fall in the category of ferromagnet (disorder), antiferromagnet, or paramagnet at all temperatures. Furthermore, the characteristic phenomena observed in spin glasses, such as the sharp 'cusp' in the frequency-dependent AC susceptibility in low fields, first observed by Cannella and Mydosh [41], is a fairly universal feature. The classical spin glass materials are noble metals (Au, Ag, Cu, etc.) weakly doped by transition metal ions (Fe, Mn, etc.). In recent years, a lot of materials have been reported in the literature, which show spin glass behavior with perovskite and other structures as discussed below. Experimentally, it has been amply demonstrated that both 3D Ising ($Fe_{0.5}Mn_{0.5}TiO_3$) and Heisenberg (Ag(Mn)) spin glass systems exhibit dynamic critical behavior on approaching the spin glass temperature, T_{sg}, which correspond to a second-order phase transition [44]. At low temperature both Ising and Heisenberg spin glasses exhibit similar non-equilibrium dynamics and an infinitely slow approach toward a thermodynamic equilibrium state. This means that below T_{sg}, the ZFC spin glasses never reach equilibrium, in other words the equilibration time is infinite for spin glass state. Therefore, the experiments on low-temperature phases seem to be like a non-equilibrium system and the results are age dependent [42]. Once the system is kept constant at low-temperature phase, it spontaneously and continuously reorganizes the spin structure, i.e., the system ages. Aging has a very good characteristic influence on the response function of spin glass. An isothermal aging experiment on an Ising and a Heisenberg system reveals that the aging has much larger influence on the relaxation of the Heisenberg system than the Ising system [45].

There are two main approaches to describe the spin glass behavior; one is phase space (mean field) and another one is the real space (droplet scaling) model. The mean field model predicts a finite spin glass temperature (e.g., T_{sg}) and also a persistence of the phase in an applied magnetic field, and the spin glass and paramagnetic phases are separated by the Almeida-Thouless line [46]. On the other hand, the droplet scaling theory predicts that in the thermodynamic limit, any finite magnetic field destroys the spin glass phase [47]. A crucial point in this model is the correspondence between the time and length scales. An experimental probe that measures at a certain frequency or timescale can also probes the system on a length scale set by the observation time (and temperature). A finite field sets an upper limit to the correlation length scales in the spin glass; on shorter length scales the system appears to be unaffected by the field, but on larger length scales the system will be at equilibrium (paramagnetic) state.

Therefore to establish the spin glass behavior experimentally, the next question that arises is what properties does a system have in order to be a spin glass? The defining properties are (i) frozen-in magnetic moments below some freezing temperature, T_{sg}, and hence a cusp in the AC susceptibility below T_{sg}; (ii) absence of periodic long-range magnetic ordering; and (iii) remanence and magnetic relaxation on macroscopic timescale below T_{sg}, at sufficiently low field. Of course, whether the moments are frozen-in or not depends on the timescale of the observation. In low-field DC magnetization, the spin glass transition, T_{sg}, is revealed by a maximum in the ZFC magnetization, irreversibility between the ZFC and the FC magnetization, and a continuous decay of the thermoremanent magnetization (TRM) to zero at the temperature where irreversibility between the ZFC and FC appears. An additional and remarkable feature of the non-equilibrium spin glass phase is "memory" phenomena. This is revealed by measurements according to a standard ZFC magnetization protocol [48]. The memory-aging behaviors are manifestations of some crucial concepts—aging, rejuvenation, and chaos—that characterize the spin glass phase and are the key factors for modeling spin glasses [45]. The frequency-dependent cusp was first reported in dilute metallic alloy of CuMn (with 0.9 % Mn) and thereafter in concentrated insulator $Eu_{1-x}Sr_xS$ [49]. In recent years, some of the perovskite manganites such as $(Tb_{0.33}La_{0.67})_{0.67}Ca_{0.33}MnO_3$, $Y_{0.7}Ca_{0.3}MnO_3$, and $Th_{0.35}Ba_{0.37}Ca_{0.28}MnO_3$ are considered to exhibit spin glass behavior at low temperatures [50]. Some of the perovskite manganites like $Nd_{0.7}Sr_{0.3}MnO_3$ and $La_{0.7-x}Y_xCa_{0.3}MnO_3$ show magnetic relaxation phenomena in the FM phase indicating magnetic frustration and disorder [51].

There is another interesting and well-known phenomenon of spin glasses called re-entrant spin glass transition, which occurs near the phase boundary between the spin glass and FM phase [42, 52]. This re-entrant spin glass behavior is reported in perovskite $Y_{0.7}Ca_{0.3}MnO_3$ [48] and in Mn-rich $YMnO_3$ hexagonal manganites [53]. Therefore, a study of re-entrant spin glass transition in a dilute magnet by Abiko et al. [54] has established a theoretical model that settles the most important issue of the re-entrant spin glass transition. Perovskite cobaltites of the type $Ln_{1-x}A_xCoO_3$ are somewhat similar in physical properties to the manganites. The spin glass behavior in $La_{1-x}Sr_xCoO_3$ has been reported by a few workers [55]. The system possesses a significant phase fraction of low-spin and SG/CG phases even in the ferromagnetic phase at high doping level, with various phases competing over the whole doping range as reported by Kuhns et al. [40]. Similarly, Burley et al. [56] and Kundu et al. [40] have reported the long-range ferromagnetism and glassy behavior in $La_{1-x}Ca_xCoO_3$ cobaltites. In some of the layered 112 perovskite cobaltites, the glassy state is formed due to the presence of EPS with the assistance of an external field, which is distinctly different from the SG state [27].

1.4.6 Multiferroicity in Perovskites

The extensive research of magnetic and ferroelectric materials has attracted increasing attention in recent years due to their possible applications toward storage devices, sensing/actuating devices, and intriguing fundamental physics [57–59]. As for the trends toward device miniaturization and high-density data storage, an integration of multifunctions into one material system has become highly desirable. The coexistence of several order parameters will bring out novel physical phenomena and offers possibilities for new devices. The novel prototype devices based on multiferroic functions may offer particularly super performance for spintronics, e.g., reading the spin states, and writing the polarization states to reverse the spin states by electric field, to overcome the high-writing energy in magnetic random-access memories.

The magnetoelectric effect, which describes the coupling between electric and magnetic fields in matters (i.e., induction of magnetization (*M*) by an electric field (*E*) or polarization (*P*) generated by a magnetic field (*H*) as shown in Fig. 1.17), is not a new phenomenon. In 1888, Röntgen observed that a moving dielectric body placed in an electric field became magnetized, which was followed by the observation of the reverse effect: polarization generation of a moving dielectric in a magnetic field [60]. Both, however, are not the intrinsic effects of matters. In 1894, by crystal symmetry consideration, Curie predicted the possibility of an intrinsic magnetoelectric effect in some crystals [61]. Subsequently, Debye coined this kind of effect as a "magneto-electric effect" [62]. The first successful observation of the magnetoelectric effect was realized in Cr_2O_3 [63]. A material that exhibits two or more primary ferroic properties such as ferromagnetism, ferroelectricity, ferroelasticity , or ferrotoroidicity is described as a multiferroic as shown in Fig. 1.17. This definition was originally proposed by Schmid in an effort to characterize materials and the effects that allow the formation of switchable domains [57–59].

The ability to combine magnetic and ferroelectric properties within one material and the potential functionality that can be achieved has resulted in much of the early work on multiferroics being concentrated within magnetic ferroelectrics [57, 64]. Among the naturally existing oxides, the presence of both ferromagnetism and

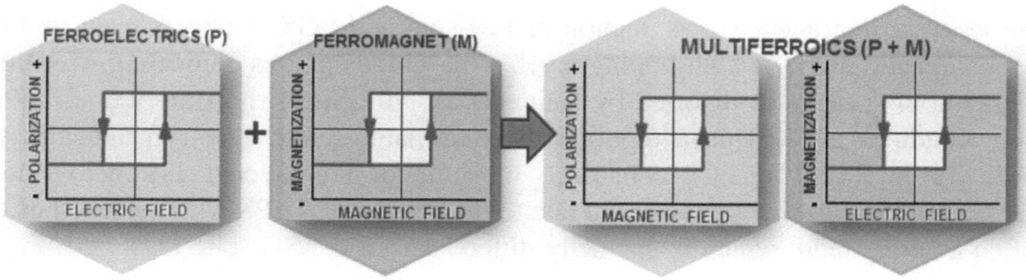

Fig. 1.17 Schematic representations of multiferroicity and the relationship between ferroelectric and ferromagnetism

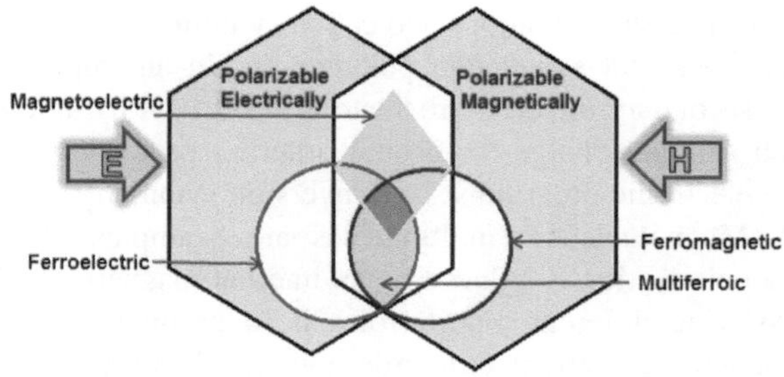

Fig. 1.18 Schematic representations of coupling between magnetic and electric degrees of freedom; it also exhibits the relationship between multiferroics and magnetoelectrics (adapted from Ref. [58])

ferroelectricity (FE) is a rare phenomenon. This phenomenon often occurs in perovskite-type oxides having the general formula ABO_3 [6–9, 65].

Coupling of the different ferroic parameters within the multiferroic system tends to be weak. The microscopic mechanisms of magnetism and ferroelectricity are very different from each other and therefore do not strongly interfere [6–9, 57, 58, 64]. The exclusivity between magnetism and ferroelectricity and the microscopic conditions required for the coupling of these different degrees of freedom can be explained using the symmetry requirements shown in Fig. 1.18 [58]. A problem when trying to design a new multiferroic material is that multiferroics do not follow one specific theory [9, 66]. The microscopic nature of magnetic ordering is well understood and generally follows the same principles as all insulating magnetic materials [6–9, 64, 65]. A material will possess a magnetic moment if it contains transition metal or rare-earth ions with partially filled *d* or *f* electron orbitals [6].

Ions with completely filled orbitals are nonmagnetic as the spins of the electrons add to zero and so do not participate in the magnetic ordering. Most of the perovskite ferroelectrics are transition metal oxides containing transition metal ions with empty *d*-orbitals. Traditionally, these materials become ferroelectric when the positively charged metal ions form covalent bonds with neighboring negatively charged oxygen ions through the virtual hopping of electrons from the filled oxygen shells to the empty *d*-orbitals. Although magnetism and ferroelectricity share the same mechanism of electron exchange, it is the contrast of empty and partially filled *d* or *f* electron shells that make the properties mutually exclusive [65]. A small group of perovskite multiferroics displays coupling between magnetic and electrical ordering. The overlap between the magnetoelectric effect and multiferroicity is not surprising as large magnetoelectric responses are expected within materials that display strong internal electromagnetic fields, often found within ferromagnets and ferroelectrics which display the largest magnetic susceptibilities and dielectric constants, respectively. Not all perovskite materials displaying the magnetoelectric effect are multiferroic, as described in the schematic diagram in Fig. 1.18. In a proper ferroelectric, polarization is a primary effect when inducing ferroelectricity; driven by hybridization and strong covalency or other purely structural effects. For example, the collective shift of anions and cations within a periodic lattice will give rise to a spontaneous and switchable polarization. In an improper ferroelectric, polarization is a secondary effect. Ferroelectricity is driven by an electronic degree of freedom such as spin, charge, or orbital ordering producing polarization as a by-product. The electronic order must lack inversion symmetry if ferroelectricity is to be induced. Magnetoelectric multiferroics are examples of improper ferroelectrics when a polarization is induced by an internal magnetic field [66]. Another method of classifying different multiferroics is to group them according to the origin of the magnetic and ferroelectric ordering: type-I and type-II. Within a type-I multiferroic, the two-order parameters have different sources. Some coupling exists but the ferroelectric ordering temperature is generally higher than the magnetic one. Ferroelectricity within a type-II multiferroic occurs as a result of magnetic ordering. Strong coupling is expected between the two parameters as ferroelectricity sets at the same temperature as magnetic ordering and is driven by it. Hence, the polarization is typically smaller within this group [9, 67].

Considering that little attention has been paid to multiferroicity until recently, it now offers us the opportunity to explore some important issues which have rarely been reachable [57–59]. Although ferroelectricity and magnetism have been the focus of condensed matter physics and materials science since their discovery, quite a number of challenges in dealing with multiferroicity within the framework of fundamental physics and technological applications have emerged. They are typically strongly correlated electronic systems in which the correlations among spins, charges/dipoles, orbitals, and lattice/phonons are significant. Therefore, intrinsic integration and strong coupling between ferroelectricity and magnetism are essentially related to the multi-latitude landscape of interactions between these orders, thus making the physics of multiferroicity extremely complicated. Nevertheless, it is also clear that multiferroicity provides a more extensive platform to explore the novel physics of strongly correlated electronic systems, in addition to high-temperature superconductor, CMR, EPS, SG, etc. in magnetic perovskites. Attempts to combine the dipole and spin orders into one system started in 1960s, [68, 69] and some multiferroics, including boracites ($Ni_3B_7O_{13}I$, $Cr_3B_7O_{13}Cl$), [69] fluorides ($BaMF_4$, M = Mn, Fe, Co, Ni), [70] $MnWO_4$, [64] $LuFe_2O_4$, [71] magnetite Fe_3O_4, [72] (Y/Yb)MnO_3, [73] $BiFeO_3$, [74] $BiMnO_3$, [75] $(La/Bi)_2MnNiO_6$, $LnMn_2O_5$, and $YBaCuFeO_5$ [6–9, 76] were identified in the following decades. However, such a combination in these multiferroics has been proven to be unexpectedly tough. Moreover, a successful combination of the two orders does not necessarily guarantee a strong magnetoelectric coupling and convenient mutual control between them. Fortunately, recent work along this line has made substantial progress by discovering/inventing some multiferroics, mainly in the category of frustrated magnets, which demonstrate the very strong and intrinsic magnetoelectric coupling [6–9, 57–59, 64, 65].

In the process of exploration of a multiferroic perovskites, the following facts are now well established: (i) FE and FM are mutually exclusive due to the d^0 electronic structure of the B-element, [6, 65] (ii) the occupation of different B-site cations with varying ionic radius provides an opportunity to realize a polar ground state, [77] and (iii) the lattice distortion induced by cations with lone pair electrons such as Pb^{2+} or Bi^{3+} plays a vital role on the FE properties as shown for perovskite $PbTiO_3$ in comparison with $BaTiO_3$ [6–9]. Bismuth-based perovskites have been recognized as materials of potential interest for their eventual multiferroic properties by the studies carried out on $BiFeO_3$ [74] and $BiMnO_3$ [75]. In these oxides, magnetism originates from superexchange interactions between iron and manganese cations through oxygen and ferroelectricity is most probably linked to the lone pair cation Bi^{3+} which induces structural distortions. Studies on perovskite $BiMn_{0.5}Ni_{0.5}O_3$ throw light on synthesis of materials with one or more order parameters for realizing multiferroic properties or magnetoelectric effects [76]. Among most of the Bi^{3+}-based perovskite systems studied for multiferroic and magnetodielectric properties, they exhibit high sensitivity toward the B-site cationic ordering, but require high pressure conditions for synthesis. Thus, the research of ferromagnetic insulators containing bismuth is of importance in order to generate new magnetoelectric properties.

References

1. G.H. Jonkar, J.H. Van Santen, Physica **16**, 377 (1950); J.H. Van Santen, G.H. Jonkar, Physica **16**, 599 (1950); G.H. Jonker, J.H. Van Santen, Physica **19**, 120 (1953); J.B. Goodenough, J. Phys. Chem. Solids **6**, 287 (1958); G.H. Jonker, J. Appl. Phys. **37**, 1424 (1966)
2. C.N.R. Rao, B. Raveau (eds.), *Colossal Magnetoresistance, Charge Ordering and related properties of manganese Oxides* (World Scientific: Singapore, 1998); Y. Tokura (eds.), *Colossal Magnetoresistance Oxides* (London: Gorden and Breach, 1999)
3. E. Dagotto (eds), *Nanoscale Phase Separation and Colossal Magnetoresistance* (Berlin: Springer, 2003); C.N.R. Rao, A.K. Kundu, M.M. Seikh, L. Sudheendra, Dalton Trans. **19**, 3003 (2004); V.B. Shenoy, C.N.R. Rao, Phil. Trans. R. Soc. A **366**, 63 (2008)
4. A.P. Ramirez, J. Phys: Condens. Matter. **9**, 8171 (1997); E. Dagotto, T. Hotta, A. Moreo, Phys. Rep. **344**, 1 (2001); H.Y. Hwang, S.W. Cheong, P.G. Radaelli, M. Marezio, B. Batlogg, Phys. Rev. Lett. **75**, 914 (1995)
5. C.N.R. Rao, J. Gopalakrishnan (eds.), *New Directions in Solid State Chemistry*, 2nd edn. (Cambridge University Press, 1997); C.N.R. Rao, B. Raveau (eds.), *Transition Metal Oxides: Structure, Properties and Synthesis of Ceramic Oxides*, 2nd edn. (Wiley-VCH, 1998); A.R. West, *Solid State Chemistry and its Applications* (John Wiley & Sons: Singapore, 2004); B. Raveau, M.M. Seikh, *Cobalt Oxides: From Crystal Chemistry to Physics* (Wiley-VCH, 2012)
6. N.A. Hill, J. Phys. Chem. B **104**, 6694 (2000)
7. D.V. Efremov, J. van den Brink, D.I. Khomskii, Nature Mater. **3**, 853 (2004); W. Prellier, M. P. Singh, P. Murugavel, J. Phys.: Condens. Matter **17**, R803 (2005); W. Eerenstein, M. Wiora, J.L. Prieto, J.F. Scott, N.D. Mathur, Nature Mater. **6**, 348 (2007); G. Catalan, J.F. Scott, Adv. Mat. **21**, 2463 (2009)
8. N. Hur, S. Park, P.A. Sharma, S. Guha, S.W. Cheong, Phys. Rev. Lett. **93**, 107207 (2004); N. Ikeda et al., Nature **436**, 1136 (2005); N.S. Rogado, J. Li, A.W. Sleight, M.A. Subramanian, Adv. Mat. **17**, 2225 (2005); I.A. Sergienko, E. Dagotto, Phys. Rev. B **73**, 094434 (2006); I.A. Sergienko, C. Sen, E. Dagotto, Phys. Rev. Lett. **97**, 227204 (2006); B. Kundys, A. Maignan, C. Simon, Appl. Phys. Lett. **94**, 072506 (2009)
9. T. Kimura, T. Goto, H. Shintani, K. Ishizaka, T. Arima, Y. Tokura, Nature (London) **426**, 55 (2003); T. Goto et al., Phys. Rev. Lett. **92**, 257201 (2004); N. Hur, S. Park, P.A. Sharma, S. Guha, S.W. Cheong, Phys. Rev. Lett. **93**, 107207 (2004); N. Ikeda et al., Nature **436**, 1136

(2005); C.R. Serrao et al., Phys. Rev. B **72**, 220101(R) (2005); O. Heyer et al., J. Phys.: Condens. Matter **18**, L471 (2006); J.R. Sahu, C.R. Serrao, N. Ray, U.V. Waghmare, C.N.R. Rao, J. Mater. Chem. **17**, 42 (2007); W. Eerenstein, M. Wiora, J.L. Prieto, J.F. Scott, N.D. Mathur, Nature Mater. **6**, 348 (2007); T. Kimura, Nature Mater. **7**, 291 (2008); G. Catalan, J.F. Scott, Adv. Mat. **21**, 2463 (2009); W. Wu, et al., Phys. Rev. Lett. **101**, 137203 (2008); C.N.R. Rao et al., J. Phys. Chem. Lett. **3**, 2237 (2012); R.D. Johnson et al., Phys. Rev. Lett. **108**, 067201 (2012); K. Singh et al., Phys. Rev. B **88**, 094438 (2013); N. Lee et al., Phys. Rev. Lett. **110**, 137203 (2013); D.K. Pratt et al., Phys. Rev. B **90**, 140401(R) (2014); R. Saha et al., Mater. Horiz. **1**, 20 (2014); T. Basu et al., Sci. Rep. **4**, 5636 (2014); A.K. Kundu, M.M. Seikh, P. Nautiyal, J. Magn. Magn. Mater. **378**, 506 (2015)

10. E.O. Wollan, W.C. Koehler, Phys. Rev. **100**, 545 (1955); W.C. Koehler, E.O. Wollan, J. Phys. Chem. Solids **2**, 100 (1957)
11. J.B.A.A. Elemans, B. Van Laar, K.R. Van Der Veen, B.O. Loopstra, J. Solid State Chem. **3**, 238 (1971); R. Mahendiran, S.K. Tiwary, A.K. Raychaudhuri, T.V. Ramakrishnan, R. Mahesh, N. Rangavittal, C.N.R. Rao, Phys. Rev. B **53**, 3348 (1996); Y. Tokura, N. Nagaosa, Science **288**, 462 (2000)
12. C. Zener, Phys. Rev. **81**, 440 (1951); C. Zener, Phys. Rev. **82**, 403 (1951); A.J. Millis, P.B. Littlewood, B.I. Shraiman, Phys. Rev. Lett. **74**, 5144 (1995); A.J. Millis, J. Appl. Phys. **81**, 5502 (1997)
13. J.B. Goodenough, Progress in Solid State Chem. **5**, 145 (1971); H.A. Kramers, Physica **1**, 191 (1934)
14. M.A. Ruderman, C. Kittel, Phys. Rev. **96**, 99 (1954); T. Kasuya, Prog. Theor. Phys. **16**, 45 (1956); K. Yoshida, Phys. Rev. **106**, 893 (1957); P.W. Anderson, H. Hesegawa, Phys. Rev. **100**, 675 (1955); J.B. Goodenough, Phys. Rev. **100**, 564 (1955); J.B. Goodenough, Phys. Rev. **124**, 373 (1961); J. Kanamori, J. Phys. Chem. Solids **10**, 87 (1959)
15. P. de Gennes, Phys. Rev. **118**, 141 (1960)
16. N.F. Mott, *Metal-Insulator Transitions* (Taylor and Francis, London, 1990); B.I. Shklovskii, A.L. Efros, *Electronic Properties of Doped Semiconductors* (Springer, Berlin, 1984)
17. M.N. Baibich, J.M. Broto, A. Fert, F. Nguyen Van Dau, F. Petroff, P. Etienne, G. Creuzet, A. Friederich, J. Chazelas, Phys. Rev. Lett. **61**, 2472 (1988)
18. R.M. Kusters, J. Singleton, D.A. Keen, R. McGreevy, W. Hayes, Physica B **155**, 362 (1989)
19. R. von Helmolt, J. Wecker, B. Holzapfel, L. Schultz, K. Samwer, Phys. Rev. Lett. **71**, 2331 (1993); K. Chahara, T. Ohno, M. Kasai, Y. Kozono, Appl. Phys. Lett. **63**, 1990 (1990); S. Jin, T.H. Tiefel, M. McCormack, R.A. Fastnacht, R. Ramesh, L.H. Chen, Science **264**, 413 (1994); S. Jin, H.M.O. Bryan, T.H. Tiefel, M. McCormack, W.W. Rhodes, Appl. Phys. Lett. **66**, 382 (1995); A. Urushibara, Y. Moritomo, T. Arima, A. Asamitsu, G. Kido, Y. Tokura, Phys. Rev. B **51**, 14103 (1995)
20. P.M. Levy, Solid State Phys. **47**, 367 (1994); P.M. Levy, S. Zhang, J. Magn. Magn. Mater. **151**, 315 (1995)
21. P. Schiffer, A.P. Ramirez, W. Bao, S.W. Cheong, Phys. Rev. Lett. **75**, 3336 (1995)
22. A. Arulraj, R. Mahesh, G.N. Subbanna, R. Mahendiran, A.K. Raychaudhuri, C.N.R. Rao, J. Solid State Chem. **127**, 87 (1996)
23. Y. Moritomo, A. Asamitu, H. Kuwahara, Y. Tokura, Nature **380**, 141 (1996); R. Mahesh, R. Mahendiran, A.K. Raychaudhuri, C.N.R. Rao, J. Solid State Chem. **122**, 448 (1996)
24. K.I. Kobayashi, T. Kimura, H. Sawada, K. Terakura, Y. Tokura, Nature **395**, 677 (1998)
25. M.A. Subhramanian, B.H. Toby, A.P. Ramirez, W.J. Marshall, A.W. Sleight, G.H. Kwei, Science **273**, 81 (1996)
26. G. Bricenco, H. Chang, X. Sun, P.G. Schultz, X.D. Xiang, Science **270**, 273 (1995); R. Mahendiran, A.K. Raychaudhuri, A. Chainani, D.D. Sarma, J. Phys.: Condens. Mater. **7**, L561 (1995); R. Mahendiran, A.K. Raychaudhuri, Phys. Rev. B **54**, 16044 (1996); H.W. Hsu et al., Mater. Sci. Eng. B **64**, 180 (1999); Z.H. Wang et al., Phys. Rev. B **60**, 14541 (1999); A. Maignan et al., Eur. Phys. J. B **13**, 41 (2000); V.G. Prokhorov et al., Phys. Rev. B. **66**, 132410 (2002); J. Wu, C. Leighton, Phys. Rev. B **67**, 174408 (2003); W. Tong et al., J. Phys.

Condens. Matter **16**, 103 (2004); R. Lengsdorf et al., Phys. Rev. B **69**, 140403 (R) (2004); W. Luo et al., J. Magn. Mater. **305**, 509 (2006)
27. C. Martin, A. Maignan, D. Pelloquin, N. Nguyen, B. Raveau, Appl. Phys. Lett. **71**, 1421 (1997); A. Maignan et al., J. Solid State Chem. **142**, 247 (1999); I.O. Troyanchuk et al., Phys. Rev. Lett. **80**, 3380 (1998); M. Respaud et al., Phys. Rev. B **64**, 214401 (2001); D. Akahoshi et al., J. Solid State Chem. **156**, 355 (2001); F. Fauth et al., Phys. Rev. B **66**, 184421 (2002); A.A. Taskin et al., Phys. Rev. Lett. **90**, 227201 (2003); D.D. Khalyavin et al., Phys. Rev. B **67**, 214421 (2003); Z.X. Zhou et al., Phys. Rev. B **70**, 24425 (2004); A.A. Taskin et al., Phys. Rev. B **71**, 134414 (2005); Z.X. Zhou et al., Phys. Rev. B **71**, 174401 (2005); V.P. Plakhty et al., Phys. Rev. B **71**, 214407 (2005); B. Raveau et al., J. Phys. Condens. Matter **18**, 10237 (2006); B. Raveau et al., Solid State Commun. **139**, 301 (2006); G. Aurelio et al., Physica B **384**, 106 (2006); G. Aurelio et al, Phys. Rev. B **76**, 214417 (2007); M García-Fernández et al., Phys. Rev. B **78**, 054424 (2008); M.M. Seikh et al., Chem. Mater. **20**, 231 (2008); M.M. Seikh et al., Solid State Commun. **149**, 697 (2009); A.K. Kundu et al., J. Phys. Condens. Matter **21**, 056007 (2009); T. Sarkar et al., Phys. Rev. B **83**, 214428 (2011); J. Wieckowski et al., Phys. Rev. B **88**, 054404 (2012); M.M. Seikh et al., J. Appl. Phys. **114**, 013902 (2013)
28. C.N.R. Rao, A. Arulraj, P.N. Santosh, A.K. Cheetham, Chem. Mat. **10**, 2714 (1998); P.M. Woodward, T. Vogt, D.E. Cox, A. Arulraj, C.N.R. Rao, P. Karen. A.K. Cheetam, Chem. Mat. **10**, 3652 (1998); A. Arulraj, P.N. Santhosh, R.S. Gopalan, A. Guha, A.K. Raychaudhuri, N. Kumar, C.N.R. Rao, J. Phys.: Condens. Mater. **10**, 8497 (1998)
29. N. Kumar, C.N.R. Rao, J. Solid State Chem. **129**, 363 (1997)
30. D.E. Cox, P.G. Radaelli, M. Marezio, S.W. Cheong, Phys. Rev. B **57**, 3305 (1998)
31. A. Arulraj, A. Biswas, A.K. Roychaudhuri, C.N.R. Rao, P.M. Woodward, T. Vogt, D.E. Cox, A.K. Cheetham, Phys. Rev. B **57**, R8115 (1996)
32. M. Uehara, S. Mori, C.H. Chen, S.W. Cheong, Nature **399**, 560 (1999); V. Podzorov, M. Uehara, M.E. Gershenson, T.Y. Koo, S.W. Cheong, Phys. Rev. B**61**, R3784 (2000); L. Sudheendra, C.N.R. Rao, J. Phys.: Condens. Mater. **15**, 3029 (2003)
33. L.M. Rodriguez-Martinez, J.P. Attfield, Phys. Rev. B **54**, R15622 (1996)
34. P.V. Vanitha, C.N.R. Rao, J. Phys.: Condens. Mater. **13**, 11707 (2003); P.V. Vanitha, A. Arulraj, P.N. Santosh, C.N.R. Rao, Chem. Mater. **12**, 1666 (2000)
35. B. Raveau, V. Caignaert, A.K. Kundu, Z. Anorg. Allg. Chem. (14 April 2015; doi:10.1002/zaac.201500088)
36. C.N.R. Rao, P.V. Vanitha, Curr. Opinion Solid State Mater. Sci. **6**, 97 (2002); C.N.R. Rao, P. V. Vanitha, A.K. Cheetham, Chem. Euro. J. **9**, 829 (2003); E. Dagotto, Science **309**, 257 (2005); V. B. Shenoy et al., Phys. Rev. Lett. **98**, 097201 (2007); J. Tao et al., Phys. Rev. Lett. **103**, 097202 (2009)
37. A. Moreo, S. Yunoki, E. Dagotto, Science **283**, 2034 (1999)
38. Ch. Renner, G. Aeppli, B.G. Kim, Y.A. Soh, S.W. Cheong, Nature **416**, 518 (2002); L. Zhang et. al., Science **298**, 805 (2002); M.R. Freeman et. al., Science **294**, 1484 (2001); J.C. Loudon et al., Nature **420**, 797 (2002); Liang et al., Nanoscale Res. Latt. **9**, 325 (2014) and references therein
39. P.M. Woodward, D.E. Cox, T. Vogt, C.N.R. Rao, A.K. Cheetham, Chem. Mat. **11**, 3528 (1999)
40. P.M. Raccah et al, Phys. Rev. **155**, 932 (1967); V.G. Bhide, D.S. Rajoria, C.N.R. Rao, G.R. Rao, V.G. Jadhao, Phys. Rev. B **12**, 2832 (1975); M.A. Senaris Rodriguez et al, J. Solid State Chem. **118**, 323 (1995); V.G. Sathe et al., J. Phys.: Condens. Mater. **8**, 3889 (1996); M. Imada et al., Rev. Mod. Phys. **70**, 1039 (1998); R. Caciuffo et al., Phys. Rev. B **59**, 1068 (1999); R. Ganguly et al., J. Phys. Condens. Matter **13**, 10911 (2001); V.G. Prokhorov et al, Phys. Rev. B **66**, 132410 (2002); P.L. Kuhns et al., Phys. Rev. Lett. **91**, 127202 (2003); R. Mahendiran et al., Phys. Rev. B **68**, 24427 (2003); L. Sudheendra et al., Ferroelectrics **306**, 227 (2004); A. Ghoshray et al., Phys. Rev. B **69**, 064424 (2004); M.J.R. Hoch et al., Phys. Rev. B **69**, 014425 (2004); S. Tsubouchi et al, Phys. Rev. B **69**, 144406 (2004); A.K. Kundu et al., J. Phys. Condens. Matter **16**, 7955 (2004); A.K. Kundu et al., Solid State Commun. **134**, 307 (2005); S.R. Giblin et al., Euro. Phys. Lett. **70**, 677 (2005); J. Wu, J.W. Lynn, C.J. Glinka, J. Burley,

H. Zheng, J.F. Mitchell, C. Leighton, Phys. Rev. Lett. **94**, 037201 (2005); D. Fuchs et al., Phys. Rev. B **71**, 92406 (2005); M.W. Haverkort et al., Phys. Rev. Lett. **97**, 176405 (2006); A. K. Kundu et al., J. Solid State Chem. **180**, 1318 (2007); J. Yu et al., Phys. Rev. B **80**, 052402 (2009); C. He et al, Phys. Rev. B **80**, 214411 (2009); D. Phelan et al, Phys. Rev. B **89**, 184427 (2014)
41. V. Cannella, J.A. Mydosh, Phys. Rev. B **6**, 4220 (1972); D. Sherrington, Kirkpatrick, Phys. Rev. Lett. **35**, 1792 (1975)
42. L. Lundgren, P. Svedlindh, P. Nordblad, O. Beckman, Phys. Rev. Lett. **51**, 911 (1983); K. Binder, A.P. Young, Rev. Mod. Phys. **58**, 801 (1986); J.A. Mydosh, *Spin Glasses: An Experimental Introduction* (Taylor and Francis, London, 1993)
43. G. Toulose, Commun. Phys. **2**, 115 (1977)
44. P. Nordblad, J. Phys.: Condens. Mater. **16**, S715 (2004)
45. P. Jonsson, R. Mathieu, P. Nordblad, H. Yoshino, H.A. Katori, A. Ito, Phys. Rev. B **70**, 174402 (2004)
46. J.R.L. de Almeida, D.J. Thouless, J. Phys. A: Math. Gen. **11**, 983 (1978)
47. D.S. Fisher, D.A. Huse, Phys. Rev. Lett. **56**, 1601 (1986)
48. R. Mathieu, P. Jonsson, D.N.H. Nam, P. Nordblad, Phys. Rev. B **63**, 092401 (2001); R. Mathieu, P. Nordblad, D.N.H. Nam, N.X. Phuc, N.V. Khiem, Phys. Rev. B **63**, 174405 (2001)
49. F. Maletta, W. Felsch, Phys. Rev. B **20**, 1245 (1979)
50. J.M. De Teresa, M.R. Ibarra, J. García, J. Blasco, C. Ritter, P.A. Algarabel, C. Marquina, A. del Moral, Phys. Rev. Lett. **76**, 3392 (1996); D. Sedmidubsky, J. Hejtmanek, M. Marysko, Z. Jirak, V. Hardy, C. Martin, J. Appl. Phys. **91**, 8260 (2001)
51. D.N.H. Nam, R. Mathieu, P. Nordblad, N.V. Khiem, N.X. Phuc, Phys. Rev. B **62**, 1027 (2000); R.S. Freitas, L. Ghivelder, F. Damay, F. Dias, L.F. Cohen, Phys. Rev. B **64**, 144404 (2001)
52. B.H. Verbeek, G.J. Nieuwenhuys, H. Stocker, J.A. Mydosh, Phys. Rev. Lett. **40**, 586 (1978)
53. W.R. Chen, F.C. Zhang, J. Miao, B. Xu, X.L. Dong, L.X. Cao, X.G. Qiu, B.R. Zhao, P. Dai, Appl. Phys. Lett. **87**, 042508 (2005)
54. S. Abiko, S. Niidera, F. Matsubara, Phys. Rev. Lett. **94**, 227202 (2005)
55. M. Itoh, I. Natori, S. Kubota, K. Matoya, J. Phys. Soc. Japan **63**, 1486 (1994); D.N.H. Nam, K. Jonason, P. Nordblad, N.V. Khiem, N.X. Phuc, Phys. Rev. B **59**, 4189 (1999)
56. J.C. Burley, J.F. Mitchell, S. Short, Phys. Rev. B **69**, 054401 (2004)
57. H. Schmid, Ferroelectrics **162**, 317 (1994); S.-W. Cheong, M. Mostovoy, Nature Mater. **6**, 13 (2007)
58. W. Eerenstein, N.D. Mathur, J.F. Scott, Nature **442**, 759 (2006)
59. M. Fiebig, J. Phys. D: Appl. Phys. **38**, R123 (2005)
60. W.C. Röntgen, Ann. Phys. **35**, 264 (1888)
61. P. Curie, J. Physique **3**, 393 (1894)
62. P. Debye, Z. Phys. **36**, 300 (1926)
63. V.J. Folen, G.T. Rado, E.W. Stalder, Phys. Rev. Lett. **6**, 607 (1961); G.T. Rado, V.J. Folen, Phys. Rev. Lett. **7**, 310 (1961)
64. D.I. Khomskii, J. Magn. Magn. Mater. **306**, 1 (2006); O. Heyer, N. Hollmann, I. Klassen, S. Jodlauk, L. Bohaty, P. Becker, J.A. Mydosh, T. Lorenz, D.I. Khomskii, J. Phys.: Condens. Matter **18**, L471 (2006)
65. N.A. Spaldin, M. Fiebig, Science **309**, 391 (2005)
66. S. Picozzi, C. Ederer, J. Phys.: Condens. Matter **21**, 303201 (2009)
67. J. van der Brink, D.I. Khomskii, J. Phys.: Condens. Matter **20**, 434217 (2008)
68. G.A. Smolenskii, A.I. Agranovskaya, S.N. Popov, V.A. Isupov, Sov. Phys. Usp. **3**, 1981 (1958)
69. E. Ascher, H. Rieder, H. Schmid, H. Stöessel, J. Appl. Phys. **37**, 1404 (1966)
70. M.D. Domenico, M. Eibschutz, H.J. Guggenheim, I. Camlibel, Solid State Commun. **7**, 1119 (1969); D.L. Fox, D.R. Tilley, J.F. Scott, H.J. Guggenheim, Phys. Rev. B **21**, 2926 (1980)
71. J.Y. Park, J.H. Park, Y.K. Jeong, H.M. Jang, Appl. Phys. Lett. **91**, 152903 (2007)
72. K. Kato, S. Iida, K. Yanai, K. Mizushima, J. Magn. Magn. Mater. **31–34**, 783 (1983)

73. I.G. Ismailza, S.A. Kizhaev, Sov. Phys. Solid State **7**, 236 (1965)
74. J.R. Teague, R. Gerson, W.J. James, Solid State Commun. **8**, 1073 (1970); J. Wang, J.B. Neaton, H. Zheng, V. Nagarajan, S.B. Ogale, B. Liu, D. Viehland, V. Vaithyanathan, D.G. Schlom, U. V. Waghmare, N.A. Spaldin, K.M. Rabe, M. Wuttig, R. Ramesh, Science **299**, 1719 (2003)
75. A.M. Santos, S. Parashar, A.R. Raju, Y.S. Zhao, A.K. Cheetham, C.N.R. Rao, Solid State Commun. **122**, 49 (2002); A. M. Santos, A.K. Cheetham, T. Atou, Y. Syono, Y. Yamaguchi, K. Ohoyama, H. Chiba, C.N.R. Rao, Phys. Rev. B **66**, 064425 (2002)
76. M. Azuma, K. Takata, T. Saito, S. Ishiwata, Y. Shimakawa, M. Takano, J. Am. Chem. Soc. **127**, 8889 (2005)
77. D.J. Singh, C.H. Park, Phys. Rev. Lett. **100**, 087601 (2008)

Chapter 2
Electronic Phase Separation and Glassy Behavior in Magnetic Perovskites

2.1 Introduction

Complex transition metal-based ABO_3 perovskites are known to exhibit compositional and electronic inhomogeneities arising from the existence of more than one phase in crystals of nominally monophasic composition, with the different phases in such materials having comparable compositions [1, 2]. Magnetic perovskites display a variety of effects due to such phase separation giving rise to novel electronic and magnetic properties. In the perovskite family the manganites became popular because of the colossal magnetoresistance (CMR) exhibited by them as discussed earlier in Section A [1–9]. CMR and related properties are generally explained on the basis of the double-exchange (DE) mechanism of the electron hopping between the $Mn^{3+}(t_{2g}^3e_g^1)$ and $Mn^{4+}(t_{2g}^3e_g^0)$ ions. Jahn–Teller (J–T) distortion associated with the Mn^{3+} ions and charge ordering (CO) of the Mn^{3+} and Mn^{4+} ions compete with DE and favor insulating behavior and antiferromagnetism [1]. CO in these materials is closely linked to the ordering of the orbitals. Typical of charge-ordered manganites are $Pr_{1-x}Ca_xMnO_3$ (x = 0.3–0.4) and $Nd_{0.5}Ca_{0.5}MnO_3$ which show CO around 250 K and antiferromagnetic ordering (A-type) at lower temperatures [4]. The CO states can be melted into metallic state by applying high magnetic fields. On the other hand, $Nd_{0.5}Sr_{0.5}MnO_3$ is ferromagnetic below room temperature and shows CO at lower temperatures ($\sim$150 K) accompanied by antiferromagnetism (CE-type). The nature of phase separation in the perovskite manganites depends on the average size of the A-site cations and the associated size disorder, carrier concentration or the composition (value of x), temperature, and other external factors such as magnetic and electric fields. Phases with different charge densities (carrier concentrations) as well as magnetic and electron transport properties coexist as carrier-rich ferromagnetic (FM) clusters or domains along with a carrier-poor antiferromagnetic (AFM) phase. Such an electronic phase separation giving rise to microscopic or mesoscopic inhomogeneous distribution of electrons results in a rich phase diagram [2]. What is noteworthy is that electronic phase separation is likely

A.K. Kundu, *Magnetic Perovskites*, Engineering Materials,
DOI 10.1007/978-81-322-2761-8_2

to be a general property of solids with correlated electrons such as the large family of transition metal oxides. There are indications that many of the unusual magnetic and electron transport properties of magnetic perovskites arise from phase separation.

The term phase separation or segregation implies the presence of at least two distinct phases in the sample, but the relative fractions may vary anywhere from a dilute regime, involving small domains of the minor phase (or clusters) in the matrix of the major phase, to a situation in which the fractions of the two phases are comparable. Thus, FM clusters present randomly in an AFM host matrix often give rise to a glassy behavior. As the FM clusters in an AFM matrix grow in size to become reasonably sized domains, due to effect of temperature, composition, or an applied magnetic field, the system acquires the characteristics of a genuine phase-separated system. In this section, we discuss electronic phase separation and associated effects in magnetic and electron transport properties in disordered perovskite manganites and cobaltites. The latter system also exhibits ferromagnetism and metallicity when the average size of the cations is sufficiently large and the size disorder is not excessive. The ferromagnetism in the perovskite cobaltites is considered to be due to Co^{3+}–O–Co^{4+} superexchange interactions.

2.2 Manganese-Centered Magnetic Perovskites

The first report on electronic phase separation (EPS) in perovskite manganites $La_{1-x}Ca_xMnO_3$ was presented by Wollen and Koehler [1]. They observed the presence of both FM and AFM peaks in the magnetic structure of $La_{1-x}Ca_xMnO_3$ by neutron scattering, and hence drew the conclusion that there is the simultaneous presence of FM and AFM phases in this material in 1950s [1]. In order to understand the magnetic and electron transport properties, CMR effect, etc. of magnetic perovskites, EPS is getting accepted as the phenomenon of importance [2–9]. However, for the nanoscopic electronic inhomogeneity in magnetic perovskites, both TEM, high-resolution TEM and scanning transmission electron microscopy (STEM) and STM can be used to find out the coexistence of nanoscopic charge-ordered (insulating) domains and the FM metallic domains, giving the local structural information at atomic level [9]. Due to the sensitivity of phase separation to magnetic fields, it is often difficult to identify EPS based on the magnetization and transport measurements. Teresa et al. [4] have reported on the experimental evidence for the existence of nanoscopic EPS in the perovskites of $(La_{1-x}A_x)_{2/3}Ca_{1/3}MnO_3$ (A = Y or Tb). In their report it was revealed that by a combination of volume thermal expansion, magnetic susceptibility, and small-angle neutron scattering measurements, there is a spontaneous formation of localized magnetic clusters with size of $\sim$1–2 nm above the ferromagnetic ordering temperature. Similarly, using small-angle magnetic neutron scattering, Mercone et al. [4] have correlated the evolution of FM phases induced by the magnetic field in a crystal of $Pr_{0.67}Ca_{0.33}MnO_3$ (Fig. 2.1). The modern technologies such as scanning

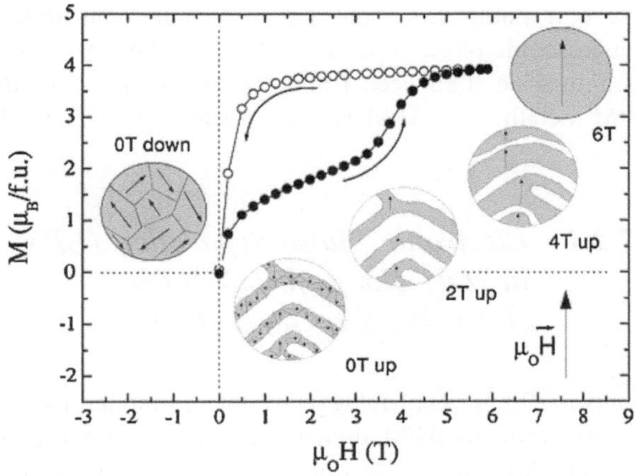

Fig. 2.1 Isotherm magnetization of $Pr_{0.67}Ca_{0.33}MnO_3$ at 30 K as a function of magnetic field and the schematic representation of EPS (adapted from Mercone et al. [4])

electron nanodiffraction, atomic-resolution STM, MFM, and electron holography have been developed that directly identify the EPS phenomenon in low-dimensional perovskite manganite nanostructures [2, 4, 6, 7]. Along with the development of nanotechnology, the EPS phenomenon in perovskite-based CMR nanoparticles has also received great attention. In recent years, the evolution of the EPS with magnetic field in perovskite nanoparticles has been reported by several groups, which has a significant impact on the perovskite-based nanoelectronics [7].

Meanwhile, Fäth et al. [7] also discovered the evidence for electronic inhomogeneities in $La_{0.7}Ca_{0.3}MnO_3$ by STM, below the FM transition temperature with a mesoscopic scale of about 0.2 μm where the FM metallic domains are interspersed in insulating regions. EPS involving submicrometer-sized FM and charge-ordered AFM domains with a typical size of about 0.2 μm was demonstrated in $La_{0.625-y}Pr_yCa_{0.375}MnO_3$ by TEM study [9]. Mesoscopic phase separation arising from the comparable energies of the ferromagnetic metallic and antiferromagnetic insulating states, with the length scale between 30 and 200 nm, is just one extreme in the perovskite manganites [9]. The EPS with phases of different charge densities is usually expected to give rise to nanometer-scale clusters as large phase-separated domains would break up into small pieces due to the Coulomb interactions. In general, microscopically homogenous clusters often ranging in their diameter size of 1–2 nm are dispersed in an insulating or charge-localized matrix. One can visualize EPS arising from disorder as well, which could arise from the size mismatch of the A-site cations in the perovskite structure [3]. Such EPS is seen in the $(La_{1-y}Pr_y)_{1-x}Ca_xMnO_3$ system in terms of a metal–insulator transition induced by disorder [9]. The size of the clusters depends on the magnitude of disorder. The smaller the disorder, the large would be the size of the clusters. This could be the reason why high magnetoresistance occurs in perovskites with small disorder.

Microscopically homogeneous clusters are usually of the size of 1–2 nm in diameter dispersed in an insulating or charge-localized matrix. Such an EPS scenario bridges the gap between the DE model and the lattice distortion models.

Several publications on the perovskite manganites reveal that in addition to microscopic phase separation, there can be mesoscopic phase separation where the length scale is between 1 and 200 nm, arising from the comparable energies of the FM metallic and AFM insulating states [1, 2, 5, 9-11].

2.2.1 Electronic Phase Separation (EPS) in $(La_{1-x}Ln_x)_{0.7}Ca_{0.3}MnO_3$ (Ln = Pr, Nd, Gd, and Y)

In $(La_{1-y}Pr_y)_{1-x}Ca_xMnO_3$ perovskites, submicrometer-sized EPS involving FM and charge-ordered AFM domains has been observed. By varying composition, the volume fraction and the domain size of the FM and charge-ordered phases could be varied [9-11]. The corresponding magnetization and resistivity data for $La_{0.5-x}Ln_{-x}Ca_{0.5}MnO_3$ (Ln = Pr, Nd) is presented in Fig. 2.2. These properties comprehensively describe the competing interactions between FM and CO/OO states and the resultant phase separation are most prominently observed.

We presented in Fig. 2.3, the temperature-dependent magnetic properties of $(La_{1-x}Nd_x)_{0.7}Ca_{0.3}MnO_3$ perovskites, which represents the sensitivity magnetic

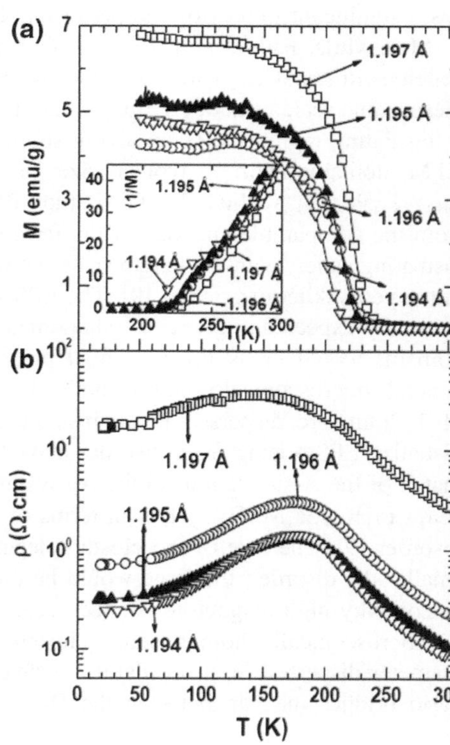

Fig. 2.2 Temperature variation of the **a** magnetization, M, and **b** resistivity, ρ, of $La_{0.5-x}Ln_xCa_{0.5}MnO_3$ (Ln = Pr, Nd) (adapted from Ref. [10]; Kundu et al.)

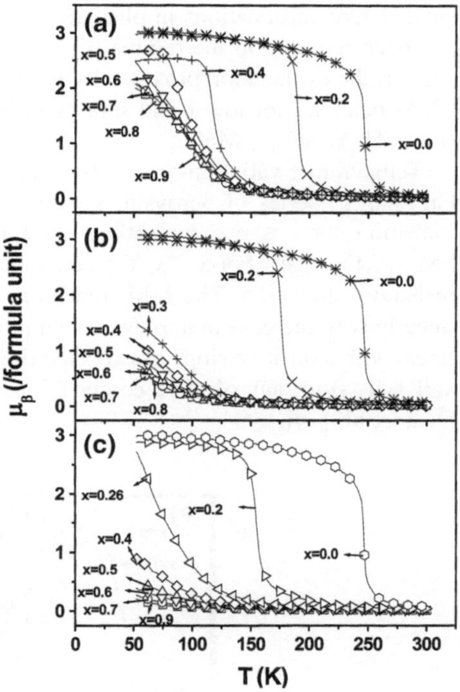

Fig. 2.3 Temperature variation of magnetic moment (μ_B) in the manganites. **a** $(La_{1-x}Nd_x)_{0.7}Ca_{0.3}MnO_3$, **b** $(La_{1-x}Gd_x)_{0.7}Ca_{0.3}MnO_3$, and **c** $(La_{1-x}Y_x)_{0.7}Ca_{0.3}MnO_3$ (adapted from Sudheendra et al. [10])

moment (μ_B) to the substitution of La by the smaller cation Nd. The FM T_C shifts to lower temperature with increase in x, a clear FM behavior is observed up to $x = 0.5$ with a saturation magnetic moment close to 3 μ_B.

Beyond $x \geq 0.6$, there is no magnetic saturation and the highest value of magnetic moment is less than 3 μ_B. The perovskite composition up to which clear FM behavior appears is defined as the critical composition x_c [10]. Whereas, the compositions with $x > x_c$ show a gradual increase in the magnetization at low temperature. In Gd-substituted perovskites, FM T_C is observed only for the doping concentration of $0.0 \leq x \leq 0.3$. The x_c ($\sim$0.3) value for Gd-substituted perovskites is much lower than the Nd perovskites ($\sim$0.6). These features show that x_c decreases with the average radius of the A-site cation, $\langle r_A \rangle$ (Fig. 2.3b), which are similar to the reported manganite perovskites by Terashita and Neumeier [12]. Similarly, Y-substituted perovskites exhibit ferromagnetism only for $x \leq 0.2$ and the FM T_C decreases with increase in x value (Fig. 2.3c). Hence for the series of $(La_{1-x}Ln_x)_{0.7}Ca_{0.3}MnO_3$ perovskite, the x_c values are 0.75, 0.6, 0.3, and 0.2 for Ln = Pr, Nd, Gd, and Y, respectively, representing a crucial dependence of x_c on $\langle r_A \rangle$. Furthermore, for these magnetic perovskites the FM is replaced with CO/OO with increase in x, which could be interpreted in terms of the change in electronic bandwidth (W). Also, the major change in the magnetization around x_c in these magnetic perovskites with constant carrier concentration could be attributed to electronic phase separation due to size disorder caused by substitution of the smaller rare earth cations in place of La. In the Pr-substituted perovskite, the EPS has been reported in the regime of $x \sim x_c$ ($x \sim 0.6$–0.8) [11]. These results also support the explanation provided by De Teresa et al. [13], where they have reported FMM behavior for low x and spin glass behavior for large x (≥ 0.33) in perovskite $(La_{1-x}Tb_x)_{0.67}Ca_{0.33}MnO_3$.

Temperature variation resistivity behavior of $(La_{1-x}Ln_x)_{0.7}Ca_{0.3}MnO_3$ (Ln = Nd, Gd, and Y) series of perovskites exhibit somewhat resemblance to the magnetic transition, the $x \leq x_c$ compositions show insulator–metal (I–M) transition near the FM T_C (Fig. 2.4). For $x > x_c$ the perovskites are insulating and do not exhibit any resistivity transition. The I–M transition, T_{IM}, for $x \leq x_c$ compositions decreases linearly with increase in doping concentration (Fig. 2.5). The T_{IM} verses $\langle r_A \rangle$ plot is linear with a positive slope as expected (inset of Fig. 2.5) and no resistivity anomaly at T ($<T_{IM}$) for any of these compositions unlike reported by Uehara et al. [9] and Deac et al. [14].

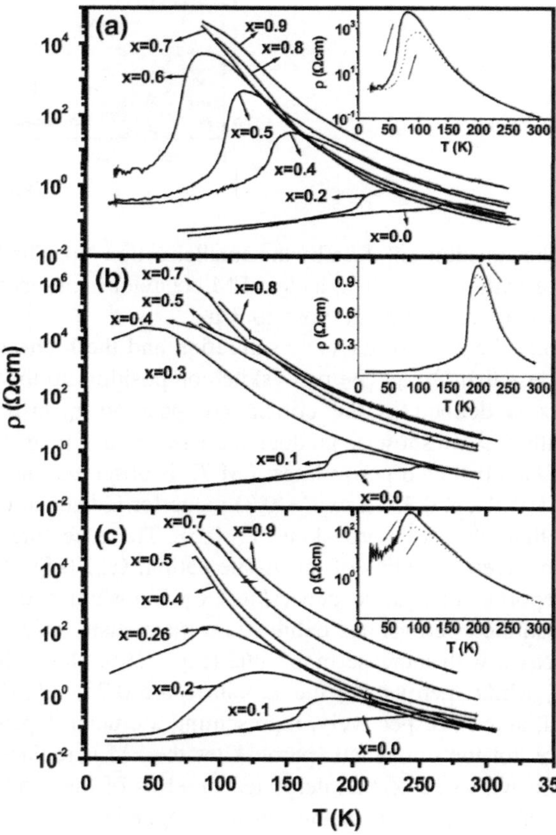

Fig. 2.4 Temperature-dependent electrical resistivity for $(La_{1-x}Ln_x)_{0.7}Ca_{0.3}MnO_3$. **a** Ln = Nd, **b** Ln = Gd, and **c** Ln = Y. (adapted from Sudheendra et al. [10])

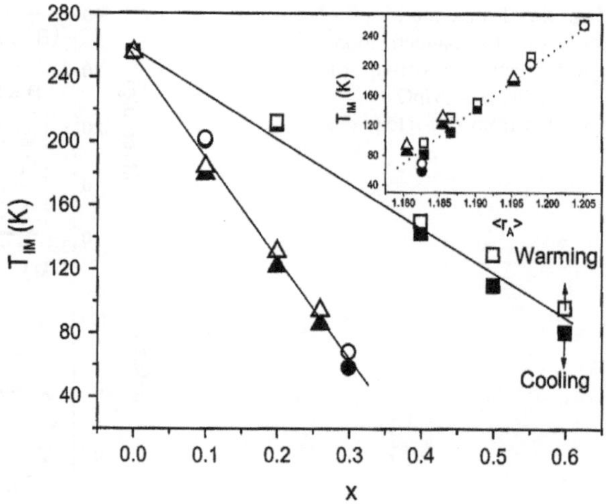

Fig. 2.5 Plot of T_{IM} versus x in $(La_{1-x}Ln_x)_{0.7}Ca_{0.3}MnO_3$ for $x \leq x_c$. The inset shows the variation of T_{IM} with $\langle r_A \rangle$ (Å) for the same composition range (adapted from Sudheendra et al. [10])

It is useful to examine how these series of perovskites leave the track on the whole magnetic and electrical behaviors and represent a case of EPS. This is corroborate from the results of small but finite magnetic moments and gradually increasing resistivities at low temperatures observed in $(La_{1-x}Ln_x)_{0.7}Ca_{0.3}MnO_3$ with Ln = Pr, Nd, Gd, and Y, around x_c or $\langle r_A^c \rangle$. These are a consequence of electronic phase separation, which also causes thermal hysteresis in the resistivity behavior around T_{IM} (insets in Fig. 2.4) [10]. The insets in Fig. 2.4a–c show that the resistivity in the warming cycle is lower than that in the cooling cycle up to a certain temperature beyond which the resistivities in the two cycles merge. With decreasing the temperature below the T_{IM}, the FMM phase grows at the expense of the antiferromagnetic insulating phase, causing a decrease in the resistivity value. Whereas during heating, the insulating phase grows at the expanse of the FMM phase, the latter provides the conductive path. The thermal hysteresis in the resistivity behavior is therefore due to the percolative conductivity in these magnetic perovskites, the hysteresis decreases with increase in $\langle r_A \rangle$ or decrease in x as expected.

2.2.2 *Electronic Phase Separation (EPS) in $(La_{1-x}Ln_x)_{0.7}(Ba/Sr)_{0.3}MnO_3$ (Ln = Pr, Nd, Gd, and Dy)*

Figure 2.6 shows the magnetization and resistivity data for $La_{0.7-x}Pr_xBa_{0.3}MnO_3$ series of magnetic perovskites [15]. The T_C value decreases progressively with increasing x and the compound is metallic at room temperature up to $x = 0.3$ and exhibits a broad I–M transition, T_{IM}, when $x \geq 0.3$. It is interesting that, as x approaches 0.7, the difference between T_C and T_{IM} increases, with the latter becoming considerably lower than T_C. $Pr_{0.7}Ba_{0.3}MnO_3$ is, therefore, a ferromagnetic insulator

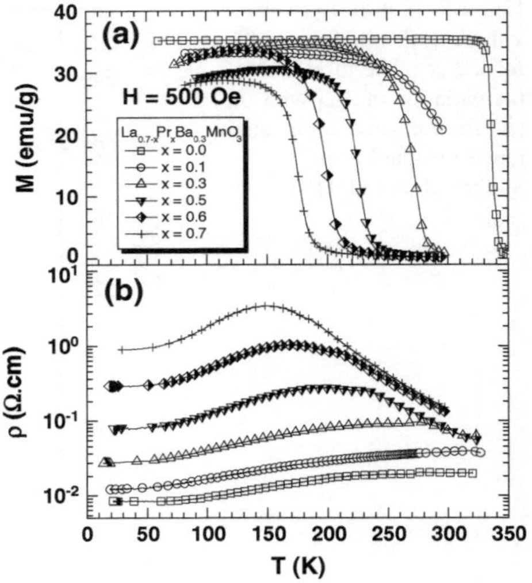

Fig. 2.6 Temperature variation of **a** magnetization and **b** electrical resistivity of $La_{0.7-x}Pr_xBa_{0.3}MnO_3$ (adapted from Ref. [15])

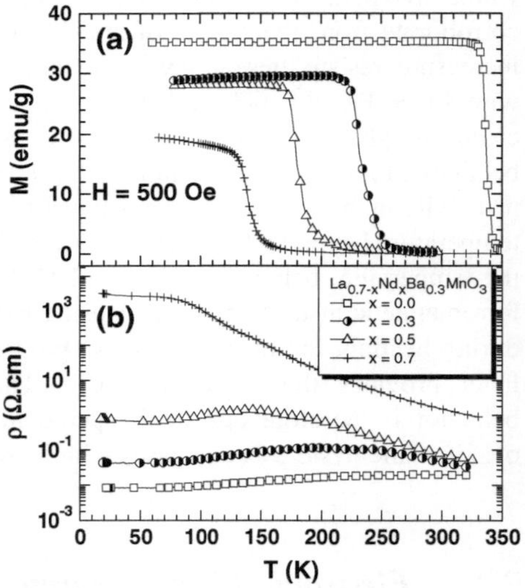

Fig. 2.7 Temperature variation of **a** magnetization and **b** electrical resistivity of $La_{0.7-x}Nd_xBa_{0.3}MnO_3$ (adapted from Ref. [15])

in the regime between T_C and T_{IM} (190–150 K). The magnetization and resistivity data of $La_{0.7-x}Nd_xBa_{0.3}MnO_3$ are shown in Fig. 2.7.

Here again, the T_C value decreases with increasing x, and there is a marked decrease in the value of the saturation magnetization as well. There is a sharp increase in magnetization with an apparent T_C of ~150 K, in $Nd_{0.7}Ba_{0.3}MnO_3$. But the saturation moment is small, suggesting that there may be no long-range ferromagnetic ordering in the compound and $Nd_{0.7}Ba_{0.3}MnO_3$ (x = 0.7) does not exhibit clear saturation down to low temperatures. The saturation magnetic

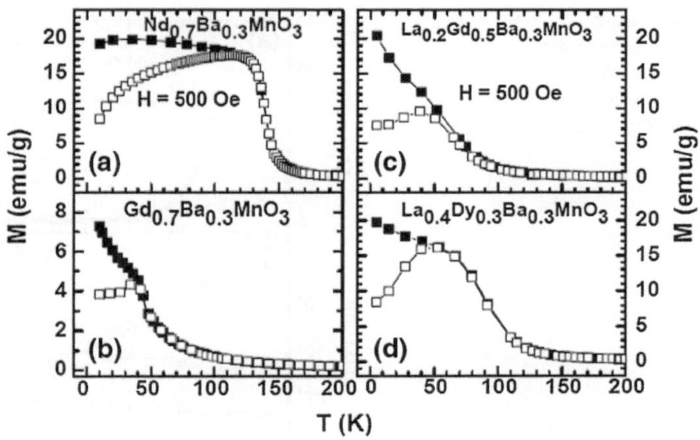

Fig. 2.8 Temperature variation of the magnetization, M, of **a** $Nd_{0.7}$ $Ba_{0.3}MnO_3$ **b** $Gd_{0.7}$ $Ba_{0.3}MnO_3$ **c** $La_{0.2}Gd_{0.5}Ba_{0.3}MnO_3$, and **d** $La_{0.4}Dy_{0.3}Ba_{0.3}MnO_3$ (at H = 500 Oe). The solid symbols represent FC and open symbols represent ZFC data, respectively (adapted from Ref. [15])

moments in $Nd_{0.7}Ba_{0.3}MnO_3$ and $La_{0.7}Ba_{0.3}MnO_3$ are 0.8 and 1.5 μ_B, respectively, while that in $Pr_{0.7}Ba_{0.3}MnO_3$ is 1.2 μ_B. Accordingly, the ZFC and FC data, show considerable divergence below T_C (Fig. 2.8a), unlike for the perovskite $Pr_{0.7}Ba_{0.3}MnO_3$. The resistivity behavior of $La_{0.7-x}Nd_xBa_{0.3}MnO_3$ is quite different from that of $La_{0.7-x}Pr_xBa_{0.3}MnO_3$. The $La_{0.7-x}Nd_xBa_{0.3}MnO_3$ compositions show a broad I–M transition when $x = 0.3$ and $x = 0.5$, but the 0.7 composition is an insulator with high resistivity.

The resistivity behavior of $Nd_{0.7}Ba_{0.3}MnO_3$ is different from the literature report [16] to some extent. Since, $Nd_{0.7}Ba_{0.3}MnO_3$ does not show long-range ferromagnetic ordering; it would appear that the material contains ferromagnetic clusters in the insulating matrix. The double peaks in resistivity data [16] or the shoulder near T_C also suggest the presence of phase separation. Ferromagnetic clusters in an insulating matrix would also be present in other compositions ($0.0 < x < 0.7$) where $T_C < T_{IM}$.

The $x \geq 0.5$ compositions show divergence between the ZFC and FC magnetization data (Fig. 2.8b, c), indicating the absence of long-range ferromagnetic ordering. AC susceptibility measurements reveal a weakly frequency-dependent peak at 50 and 40 K, respectively, in the $x = 0.5$ and 0.7 compositions. These compositions also fail to show the I–M transitions in the resistivity data, whereas the samples with $x < 0.5$ show distinct I–M transitions. The compositions with $x > 0.5$ are insulating similar to $Nd_{0.7}Ba_{0.3}MnO_3$ and the resistivity of $Gd_{0.7}Ba_{0.3}MnO_3$ is higher than that of $Nd_{0.7}Ba_{0.3}MnO_3$. In the $La_{0.7-x}Dy_xBa_{0.3}MnO_3$ series, ferromagnetism does not occur for $x > 0.2$ (Fig. 2.9a). The $x = 0.2$ composition shows an apparent T_C of 180 K, but the saturation magnetization is very low (18 emu/g). The $x = 0.2$ composition shows the I–M transition, but all the compositions with $x > 0.2$ are insulating, the resistivity being higher than

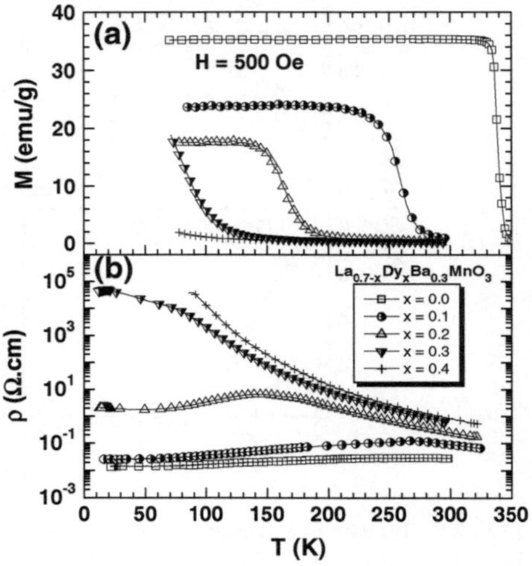

Fig. 2.9 Temperature variation of **a** magnetization and **b** electrical resistivity of $La_{0.7-x}Dy_x\ Ba_{0.3}MnO_3$ (adapted from Ref. [17])

that of the corresponding Gd- and Nd-substituted magnetic perovskites. The ZFC and FC data of the $x = 0.3$ composition shows divergence (Fig. 2.8d), indicating the absence of long-range ferromagnetic ordering.

In both the $La_{0.7-x}Gd_xBa_{0.3}MnO_3$ and $La_{0.7-x}Dy_xBa_{0.3}MnO_3$ series of magnetic perovskites, ferromagnetism disappears as x increases, accompanied by an insulating behavior. The apparent ferromagnetic transitions with a low saturation magnetization observed for $x = 0.3$ and 0.2 at 150 and 180 K, respectively, in the Gd and Dy derivatives, and associated with T_{IM} values lower than T_C, point to the presence of a ferromagnetic insulating state. It is likely that in all the compositions where the ferromagnetic insulating state occurs, there is phase separation wherein ferromagnetic clusters are present in an insulating matrix. It is interesting that the difference between T_C and T_{IM} manifests itself only when σ^2 is considerably large. In $La_{0.7-x}Ln_xBa_{0.3}MnO_3$ series, the difference between T_C and T_{IM} starts emerging when the $\sigma^2 = 0.016$ Å^2, although the $\langle r_A \rangle$ is relatively large, being around 1.28 Å. Clearly, the size disorder plays a crucial role in determining the properties of these magnetic perovskites. The effect of size disorder with constant $\langle r_A \rangle$ values corresponding to $Pr_{0.7}Ba_{0.3}MnO_3$ and $Gd_{0.7}Ba_{0.3}MnO_3$, respectively, are shown in Figs. 2.10 and 2.11. The T_C increases with decreasing σ^2 and the material becomes metallic at the lowest value of $\sigma^2 = 0.008$ Å^2, while I–M transitions occur in the σ^2 range of 0.02–0.01 Å^2. This is indeed a nice result in that a system normally showing an I–M transition becomes metallic as the size disorder is decreased. The effect of size disorder is seen more vividly when the $\langle r_A \rangle$ value is 1.216 Å, corresponding to $Gd_{0.7}Ba_{0.3}MnO_3$, a nonmagnetic insulating material. However, when the size disorder is decreased, the material becomes ferromagnetic, with the T_C going up to ~300 K at the lowest value of σ^2 (Fig. 2.11a). As σ^2 decreases, the insulating behavior also gives way to metallic behavior.

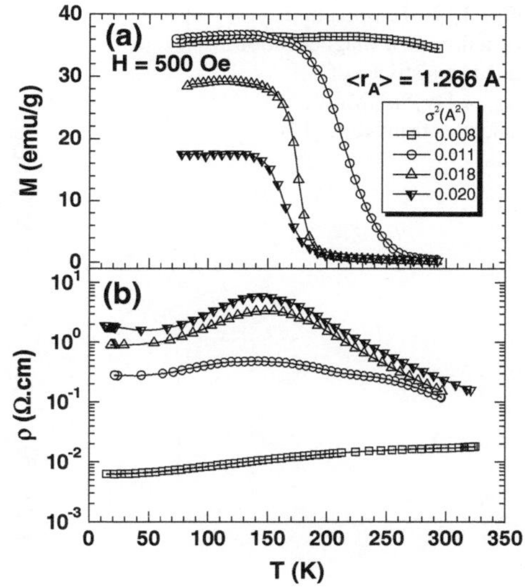

Fig. 2.10 Temperature variation of **a** magnetization and **b** the electrical resistivity of $Ln_{0.7-x}Ln'_xA_{0.3-y}A'_yMnO_3$ with a fixed $\langle r_A \rangle$ value of 1.266 Å (adapted from Ref. [15])

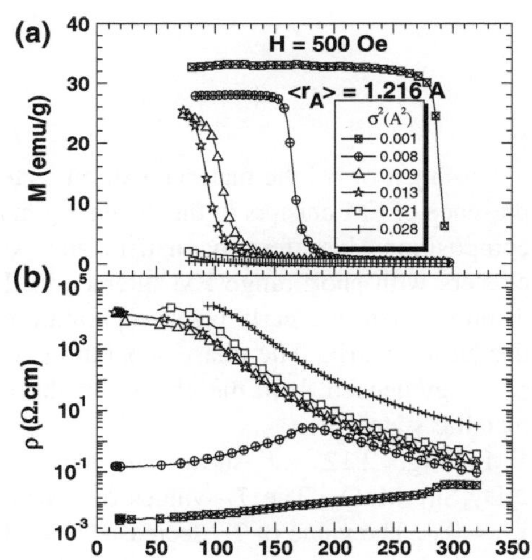

Fig. 2.11 Temperature variation of **a** magnetization and **b** the electrical resistivity of $Ln_{0.7-x}Ln'_xA_{0.3-y}A'_yMnO_3$ with a fixed $\langle r_A \rangle$ value of 1.216 Å (adapted from Ref. [15])

The electronic and magnetic properties of $La_{0.7-x}Ln_xBa_{0.3}MnO_3$ (Ln = Pr, Nd, Gd, and Dy) magnetic perovskites have revealed certain interesting aspects, wherein the average radius of the A-site cation generally remains large (1.216–1.292 Å), but the size disorder is considerable. Since the band narrowing due to small $\langle r_A \rangle$ is avoided, the predominant effect in these materials is due to size disorder. It is interesting that these materials show a progressive decrease in the FM T_C, eventually giving rise to a non-FM insulating behavior. Accordingly, with

Fig. 2.12 Temperature variation of **a** magnetization and **b** resistivity of $La_{0.7-x}Dy_xSr_{0.3}MnO_3$ (adapted from Ref. [15])

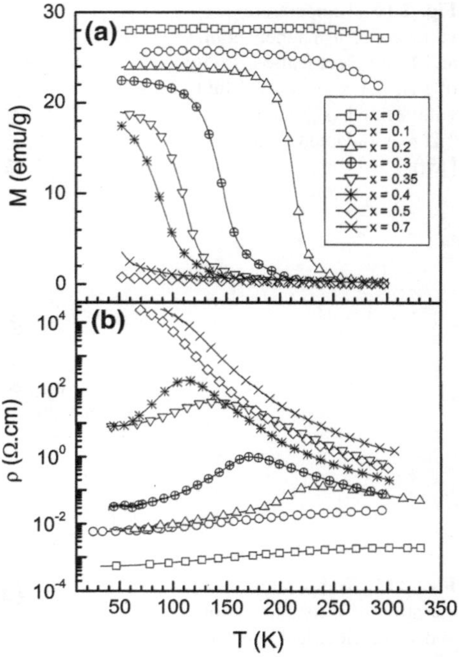

increasing x or σ^2, the material exhibits a ferromagnetic insulating phase due to the presence of FM clusters in the insulating matrix. At large x or σ^2, where some of the compositions lose ferromagnetism and become insulating, there is evidence for clusters with short-range FM interaction. In the insulating regime caused by size disorder, there is clearly phase separation due to the presence of FM clusters in an insulating matrix. The phase separation is minimized or eliminated by decreasing σ^2, as evidenced from the change of the nonmagnetic insulating phase to an FM metallic state.

In Fig. 2.12 we show the magnetization and resistivity data of $La_{0.7-x}Dy_xSr_{0.3}MnO_3$. The T_C values decrease with increasing x up to a composition $x_c \sim 0.4$. The value of T_C decreases from 350 K for $x = 0.0$ to ~ 110 K for $x = 0.4$. The abrupt change in magnetization of $La_{0.7-x}Dy_xSr_{0.3}MnO_3$ is noteworthy. There is a small increase in the magnetization at low temperatures (≤80 K) in the $x > x_c$ compositions (Fig. 2.12a), but this is not due to long-range FM ordering. If these compositions were FM the T_C value would be expected much higher based on the $\langle r_A \rangle$ value. When $x > x_c$, the materials are no longer FM and accordingly, the resistivity increases with the decrease in temperature, as in insulator (Fig. 2.12b). At large x ($x > x_c$) $La_{0.7-x}Dy_xSr_{0.3}MnO_3$ ceases to exhibit ferromagnetism and I–M transition, and instead becomes an insulator with a small increase in magnetization at low temperature indicating that the FM clusters occur in a paramagnetic matrix. The large change in the properties around x_c reflects the presence of electronic phase separation in the Sr-substituted magnetic perovskites as well.

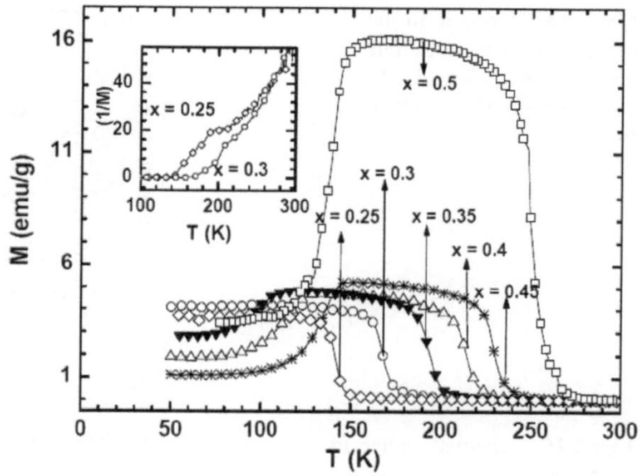

Fig. 2.13 Temperature variation of magnetization of $Nd_{0.5}Ca_{0.5-x}Sr_xMnO_3$. The *inset* shows the variation of inverse magnetization with temperature. adapted from Ref. [17]

2.2.3 *Electronic Phase Separation (EPS) in $Nd_{0.5}Ca_{0.5-x}Sr_xMnO_3$ (X = 0–0.5)*

The $Nd_{0.5}Ca_{0.5-x}Sr_xMnO_3$ perovskites [17] show the well-known ferromagnetic transition around 250 K and undergoe charge-ordering transition on cooling at 150 K when $x = 0.5$, the material becoming antiferromagnetic around the same temperature (Fig. 2.13). Both the ferromagnetic and charge-ordering transitions are sharp in this composition. The $x = 0$ composition ($Nd_{0.5}Ca_{0.5}MnO_3$) shows only charge ordering ($T_{CO} = 240$ K) but no ferromagnetism. The $Nd_{0.5}Ca_{0.5-x}Sr_xMnO_3$, compositions with $x = 0.25$–0.45 show ferromagnetic transitions (Fig. 2.13), with the T_C increasing with increase in x and the $x = 0.25$ composition has a T_C close to 150 K. However, it is noteworthy that when $x < 0.35$, there is no sharp drop in magnetization data corresponding to the charge-ordering transition. The compositions with $x = 0.25$ and 0.30 are more like $Nd_{0.5}Ca_{0.5}MnO_3$ and the temperature variation of the inverse magnetization shows a dip corresponding to T_{CO} (inset Fig. 2.13).

The occurrence of a phase-separated state below T_{CO} (T_N) in some of the manganate compositions was pointed out earlier. The situation is even more complex in $Nd_{0.5}Sr_{0.5}MnO_3$. High-resolution X-ray and neutron diffraction investigations show that $Nd_{0.5}Sr_{0.5}MnO_3$ separates into three macroscopic phases at low temperatures [17]. The phases involved are the high-temperature FMM phase, the orbitally ordered A-type AFM phase, and the charge-ordered CE-type AFM phase. On cooling the manganite, the A-type AFM phase starts manifesting itself around 220 K, with the charge-ordered AFM phases appearing at 150 K. At the so-called FM metallic-CO AFM transition, all the three phases coexist, and this situation continues down to very low temperatures as shown in Fig. 2.14.

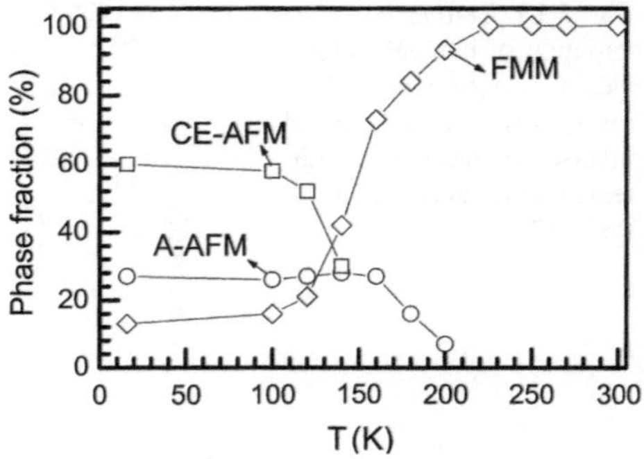

Fig. 2.14 Variation in the percentage of the different phases of $Nd_{0.5}Sr_{0.5}MnO_3$ with temperature (adapted from Ref. [17])

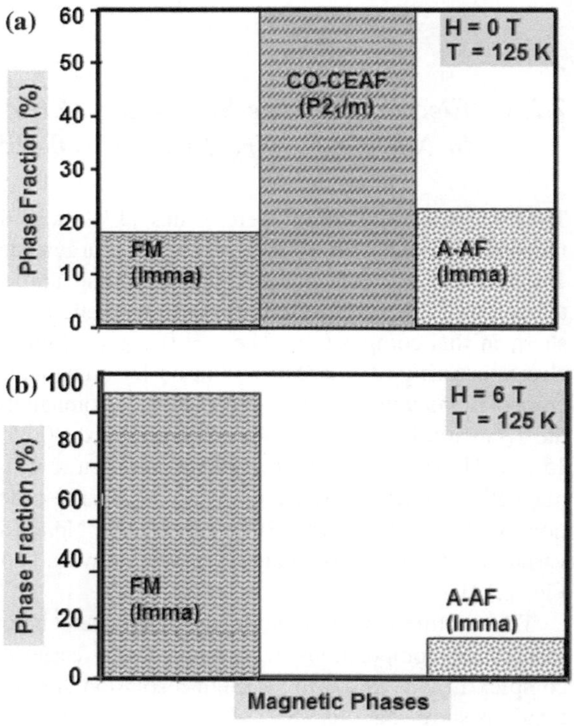

Fig. 2.15 Schematic diagram of the percentage volume fractions of different phases of $Nd_{0.5}Sr_{0.5}MnO_3$ under **a** H = 0 T and **b** 6 T (adapted from Ref. [18])

Figure 2.15 shows the percentage volume fraction of the different phases in the presence and absence of a magnetic field [18]. Phase separation in this system seems to depend crucially on the Mn^{4+}/Mn^{3+} ratio, a ratio slightly greater than unity stabilizes the A-type AFM phases. Thus, $Nd_{0.45}Sr_{0.55}MnO_3$ has the A-type AFM structure.

Figure 2.16 shows the T_C and T_{CO} values in the $Nd_{0.5}Ca_{0.5-x}Sr_xMnO_3$ and $La_{0.5-x}Ln_xCa_{0.5}MnO_3$ (Ln = Pr, Nd) series against $\langle r_A \rangle$. Although there is some scatter in the points, the data indicate that when $\langle r_A \rangle \sim 1.20$ Å, the $T_C < T_{CO}$,

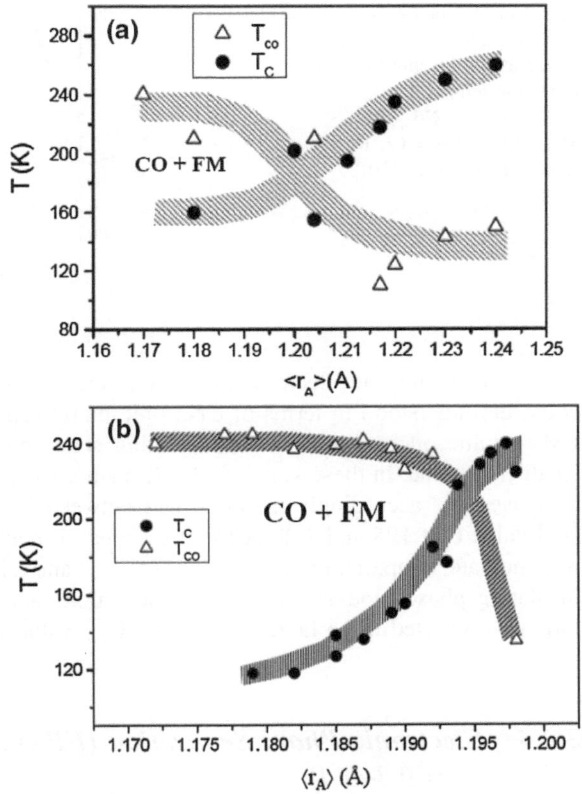

Fig. 2.16 Variation of the FM Curie temperature, T_C, and the charge-ordering transition temperature, T_{CO}, with $\langle r_A \rangle$ in **a** $Nd_{0.5}Ca_{0.5-x}Sr_xMnO_3$ and **b** $La_{0.5-x}Ln_xCa_{0.5}MnO_3$ (Ln = Pr, Nd). In the temperature range between T_{CO} and T_C, the charge-ordered and FM phases coexist (adapted from Ref. [19])

suggesting that the ferromagnetic transition is reentrant in nature [19]. Furthermore, the T_C–$\langle r_A \rangle$ and T_{CO}–$\langle r_A \rangle$ curves cross each other around $\langle r_A \rangle = 1.20$ Å. It is likely that over the entire $\langle r_A \rangle$ range 1.17–1.24 Å, there is coexistence of the CO and FM phases, especially in the temperature range between T_{CO} and T_C. In Fig. 2.17, we have shown the data for $Nd_{0.5}Ca_{0.5}Mn_{1-x}M_xO_3$ (M = Cr, Ru) series, where $\langle r_A \rangle$ is constant (1.17 Å) [20]. The T_{CO} generally decreases with increase in x, while T_C increases specially in the case of Ru substitution. It appears that this ferromagnetic transition is reentrant in nature in these magnetic perovskites and suggests that T_C and T_{CO} curves cross each other at a specific value of x. In the temperature range between T_{CO} and T_C, the charge-ordered and FM phases coexist.

In perovskite manganites, the reason behind the value $T_{CO} > T_C$ with small $\langle r_A \rangle$ ($\langle r_A \rangle < 1.20$ Å) is probably because of electronic phase separation. It is known that only when $\langle r_A \rangle < 1.20$ or 1.19 Å, charge ordering and associated effects occur in the perovskite manganites [9, 10]. The so-called FM transition is thus a consequence of electronic phase separation. The T_C's at small $\langle r_A \rangle$ do not really correspond to genuine FM transitions, and accordingly the saturation magnetization values of these samples at low temperatures are small. It is only at large $\langle r_A \rangle$, ($\langle r_A \rangle \geq 1.20$ Å) that genuine FM phases associated with high T_C values get manifested.

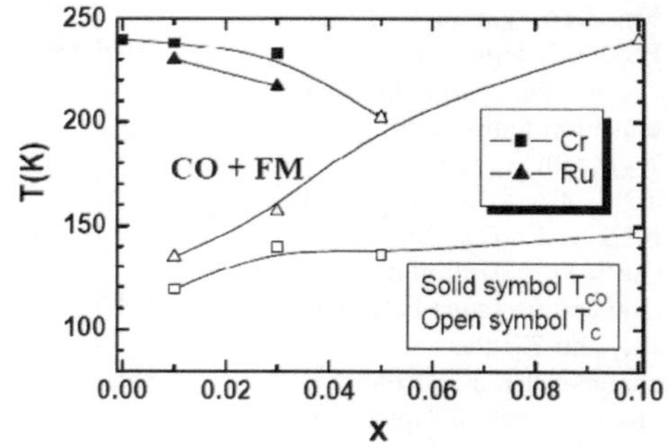

Fig. 2.17 Variation of ferromagnetic Curie temperature, T_C, and the CO transition temperature, T_{CO}, with x in $Nd_{0.5}Ca_{0.5}Mn_{1-x}M_xO_3$ (M = Cr, Ru) (adapted from Ref. [20])

The magnetic and electron transport properties of these magnetic perovskites could be understood in terms of electronic phase separation. The system becomes FM metallic on cooling the CO insulating state and this behavior is favored for small $\langle r_A \rangle$ value. In these systems, the T_C increases with increase in $\langle r_A \rangle$ while T_{CO} decreases and accordingly the crossover between T_{CO} and T_C takes place around $\langle r_A \rangle$ value of 1.195 and 1.20 Å for two series of manganites. It appears that in the intermediate temperature range between T_{CO} and T_C, the FM metallic and CO insulating phases coexist for these compositions and such electronic phase separation is expected to be favored by small $\langle r_A \rangle$ values.

2.2.4 *Electronic Phase Separation (EPS) in $Pr_{1-x}Ca_xMnO_3$ (X = 0.3–0.4)*

The studies of neutron scattering and diffraction by Radaelli et al. [4, 5] have shown tunable mesoscopic phase separation in $Pr_{0.7}Ca_{0.3}MnO_3$. Intragranular strain-driven mesoscopic phase separation (5–20 nm) between two insulating phases (one charge-ordered and another spin glass) occurs below T_{CO}. The charge-ordered phase orders antiferromagnetically and the other remains a spin glass. On the application of a high magnetic field, most of the material goes to a FM state. Microscopic phase separation (0.5–2 nm) is present at all temperatures, especially in the spin glass phase at low temperatures. Electric fields produce interesting effects on $Pr_{0.6}Ca_{0.4}MnO_3$ and similar CO magnetic perovskites.

In Fig. 2.18, we show the effect of electric currents on the resistivity of $Pr_{0.6}Ca_{0.4}MnO_3$ and $Nd_{0.5}Ca_{0.5}MnO_3$ crystals when the sample is cool down to 15 K from 300 K. There are four distinct features in the plots. There is a drop in the resistivity throughout the temperature range as the current, I, is increased. The temperature dependence of resistivity changes with the increase in I. An insulator–metal transition occurs around 60 K (T_{IM}) at high values of I, beyond a threshold value. The change in the resistivity is not due to Joule heating as evidenced from the

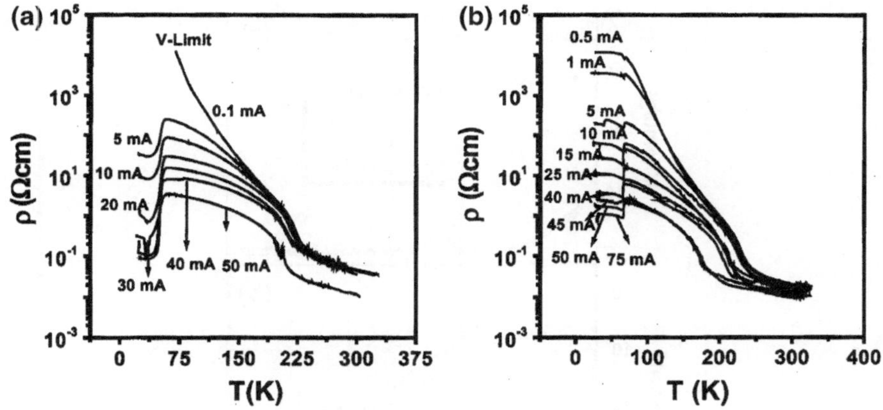

Fig. 2.18 Temperature variation of electrical resistivity of **a** $Pr_{0.6}Ca_{0.4}MnO_3$ and **b** $Nd_{0.5}Ca_{0.5}MnO_3$ for different values of current (adapted from Rao et al. [8])

negative thermal coefficient of resistivity at high temperatures and the change in sign at the I–M transition [8]. The negative differential resistance, i.e., the decrease in resistivity with increase in current observed beyond a certain value of I (Fig. 2.18), is due to the presence of the metallic filaments, which are ferromagnetic and carry most of the current [8]. The rather high value of the resistivity below the transition temperature is attributed to the coexistence of the FM and CO insulator phases. The relative fraction of the FMM phase increases with increasing current causing a lowering of resistivity below the T_{IM} [8]. The small rise in resistivity below the I–M transition may be attributed to the tunneling of electrons between the FMM clusters through COI clusters.

2.2.5 *Glassy Ferromagnetism in $Ln_{0.7}Ba_{0.3}MnO_3$ (Ln = La, Nd, and Gd)*

We have discussed earlier that the perovskite manganites $(La_{1-x}Tb_x)_{0.67}Ca_{0.33}MnO_3$ is FMM for lower doping x and spin glass at large x (≥0.33). In this section we will discuss for $Ln_{0.7}Ba_{0.3}MnO_3$ manganites in this direction. Figure 2.19 shows the magnetization and resistivity behavior of $La_{0.7}Ba_{0.3}MnO_3$, $Nd_{0.7}Ba_{0.3}MnO_3$ and $Gd_{0.7}Ba_{0.3}MnO_3$ [15]. Clearly, the magnetic properties of the three perovskites are distinctly different from one another. $La_{0.7}Ba_{0.3}MnO_3$ shows FMM behavior below T_C, whereas $Nd_{0.7}Ba_{0.3}MnO_3$ and $Gd_{0.7}Ba_{0.3}MnO_3$ show insulating behavior with complex magnetism. Figure 2.20 shows the low-field ZFC and FC magnetization data of $Nd_{0.7}Ba_{0.3}MnO_3$ and $Gd_{0.7}Ba_{0.3}MnO_3$, the later one $Gd_{0.7}Ba_{0.3}MnO_3$ exhibits a rather complex behavior below 62 K.

It is noteworthy that there are three characteristic temperatures: 62 K (onset of significant irreversibility between the ZFC and FC magnetization curves), 46 K

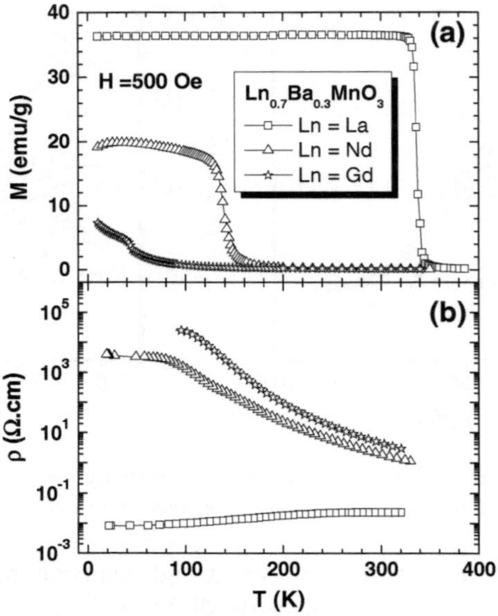

Fig. 2.19 Temperature dependence of **a** the FC magnetization, M, (at H = 500 Oe) and **b** the electrical resistivity, ρ, of $Ln_{0.7}Ba_{0.3}MnO_3$ with Ln = La, Nd, and Gd. Note that $Nd_{0.7}Ba_{0.3}MnO_3$ is insulating at 150 K where there is weak magnetic transition (adapted from Ref. [15])

(a maximum in the FC curve), and 36 K (a maximum in the ZFC curve), all indicating different ordering and/or freezing processes in the system. The field variation of magnetization at different temperatures for $Nd_{0.7}Ba_{0.3}MnO_3$ shows a behavior similar to a soft ferromagnet. However, $Gd_{0.7}Ba_{0.3}MnO_3$ does not show a ferromagnet at low temperatures, and exhibit no saturation even at high fields. The shape of the M-H curve and the absence of saturation even at high fields found in $Gd_{0.7}Ba_{0.3}MnO_3$ are reminiscent of magnetization curves of spin glasses [21].

Figure 2.21 shows the in-phase $\chi'(T)$ component of the AC susceptibility, which reveals similar features as the ZFC magnetization at low field in both the manganites. $Nd_{0.7}Ba_{0.3}MnO_3$ shows a sharp frequency-independent maximum below 150 K. However, there is weak frequency dependence at temperatures below 140 K, a behavior similar to $Nd_{0.7}Sr_{0.3}MnO_3$ [22]. $Gd_{0.7}Ba_{0.3}MnO_3$ shows a shoulder around 62 K, a weak anomaly just above 46 K and a maximum at 36 K. The $\chi'(T)$ data become strongly frequency dependent below 36 K. This transition could arise from the presence of small magnetic clusters in a nonmagnetic matrix. Time-dependent ZFC magnetization measurements show that both $Nd_{0.7}Ba_{0.3}MnO_3$ and $Gd_{0.7}Ba_{0.3}MnO_3$ exhibit logarithmically slow dynamics and aging at low temperatures (Fig. 2.22).

Magnetic aging is a signature of spin glasses and explained within the droplet (or domain growth) model the maximum in the relaxation rate is associated with a crossover between quasi-equilibrium and nonequilibrium dynamics. The slow

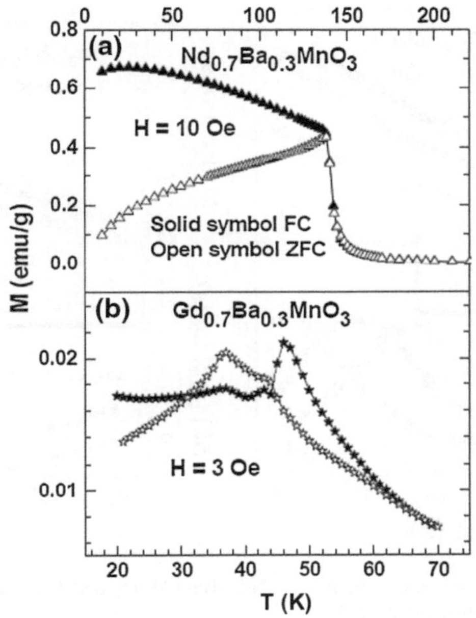

Fig. 2.20 Temperature-dependent magnetization for **a** $Nd_{0.7}Ba_{0.3}MnO_3$ (at H = 10 Oe) and **b** $Gd_{0.7}Ba_{0.3}MnO_3$ (at H = 3 Oe) (adapted from Ref. [15])

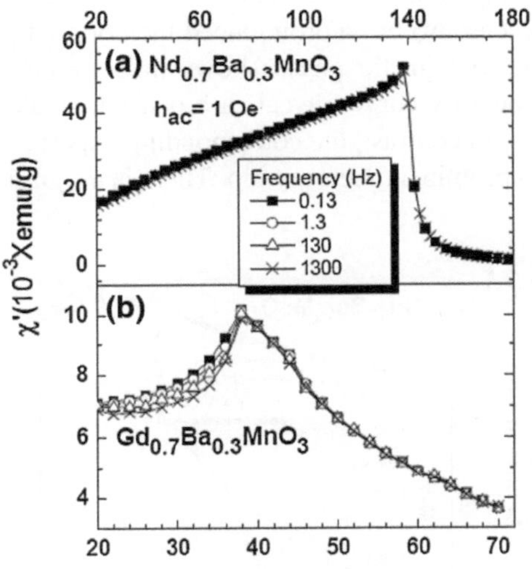

Fig. 2.21 Temperature dependence of the in-phase AC susceptibility at different frequencies for **a** $Nd_{0.7}Ba_{0.3}MnO_3$ and **b** $Gd_{0.7}Ba_{0.3}MnO_3$ (adapted from Ref. [15])

relaxation and aging behavior of $Nd_{0.7}Ba_{0.3}MnO_3$ and $Gd_{0.7}Ba_{0.3}MnO_3$ demonstrate that magnetic disorder and frustration occur in the low temperature phases. Glassy dynamics in spin glasses is also manifested by a memory effect that can be

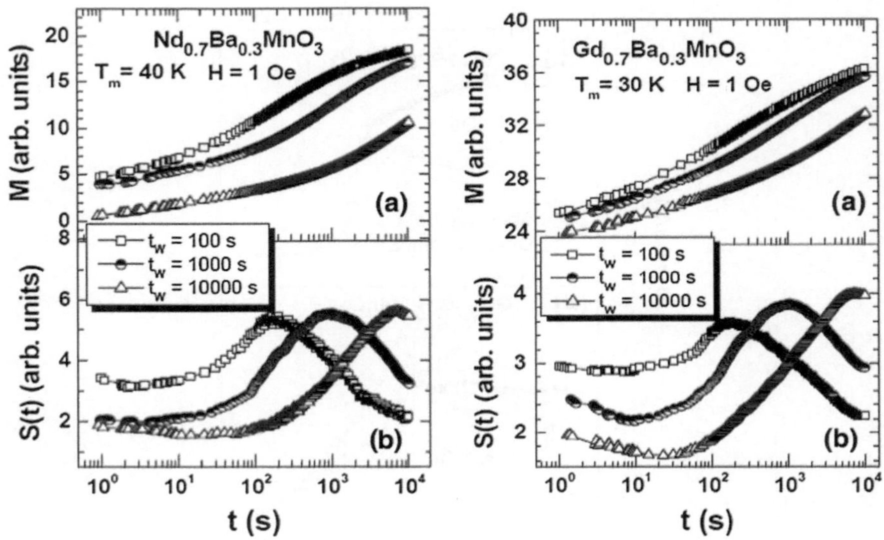

Fig. 2.22 ZFC-relaxation measurements for $Nd_{0.7}Ba_{0.3}MnO_3$ and $Gd_{0.7}Ba_{0.3}MnO_3$ for different waiting times (adapted from Ref. [15])

demonstrated by DC magnetization or low frequency AC susceptibility experiments. In a spin glass experiment, the memory curve acquires a weak dip at the temperature where the zero-field cooling was halted. A spin glass phase (ordinary or reentrant) has a pronounced memory behavior, whereas a disordered and frustrated ferromagnetic phase shows little or no memory effect. Figure 2.23a shows the memory curve, reveals a broad but shallow memory of the stop at 120 K.

In contrast, the corresponding experiment on the $Gd_{0.7}Ba_{0.3}MnO_3$ sample shows a prominent memory dip. There is a significant difference between the reference and

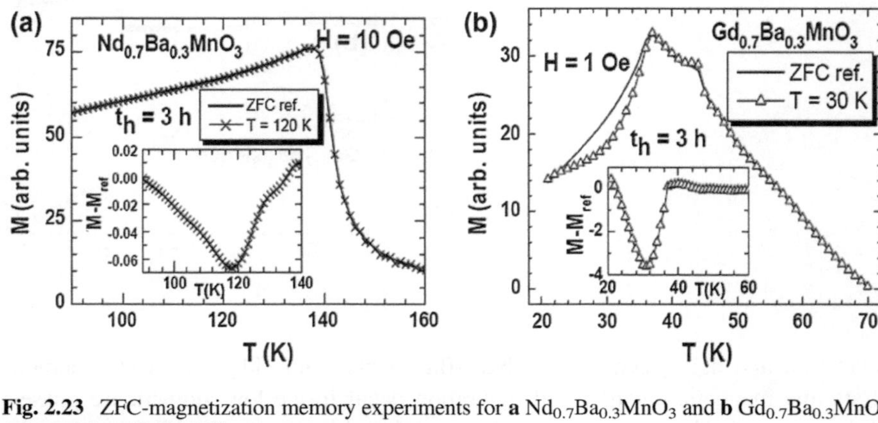

Fig. 2.23 ZFC-magnetization memory experiments for **a** $Nd_{0.7}Ba_{0.3}MnO_3$ and **b** $Gd_{0.7}Ba_{0.3}MnO_3$ (adapted from Ref. [15])

the memory curves, with a broad memory dip. The dip abruptly ceases above 36 K, hence the memory behavior of $Gd_{0.7}Ba_{0.3}MnO_3$ at 36 K signifies spin glass behavior. $Nd_{0.7}Ba_{0.3}MnO_3$ shows a pronounced aging behavior, but a rather weak memory effect below 150 K, probably due to the presence of FM clusters in an insulating matrix. It appears to be a cluster glass or a magnetically disordered system. $Gd_{0.7}Ba_{0.3}MnO_3$ appears to contain small magnetic clusters, giving rise to a spin glass state below 36 K. The behavior of $Gd_{0.7}Ba_{0.3}MnO_3$ is attributed to the large size mismatch between the A-site cations or large σ^2 value (0.028 Å^2), the mismatch being considerably smaller in $Nd_{0.7}Ba_{0.3}MnO_3$. Such size mismatch favors chemical/electronic inhomogeneities, this is a unique case of a perovskite manganite showing a size disorder-induced spin glass behavior, occurring in spite of the relatively large A-site cation radius ($\langle r_A \rangle$ = 1.216 Å). The FM insulating state or nonmagnetic insulating state often reported in the rare earth manganites of the type $Ln_{1-x}A_xMnO_3$ arises from the glassy behavior of the magnetic clusters in these materials, generally associated with electronic phase separation [2–8].

2.3 Cobalt-Centered Magnetic Perovskites

The physical properties of perovskite cobaltites $Ln_{1-x}A_xCoO_3$ (Ln = rare earth, A = alkaline earth) are somewhat similar to that of the perovskite manganites, unlike the spin state transitions in the case of perovskite cobaltites [23, 24]. One of the fundamental aspects of the transition metal chemistry is ligand field theory, which provides the possible explanation for the spin state transitions of perovskite cobaltites. In this section we have discussed the effect of A-site cation and size disorder on the magnetic and electrical properties as well as the factors that influence the EPS and spin glass behavior. Recently, Phelan et al. [24] have reported the EPS and MR effect in perovskite cobaltites $Pr_{0.5}Ca_{0.5}CoO_{3-\delta}$ and $(Pr_{1-y}Y_y)_{1-x}Ca_xCoO_{3-\delta}$, where the ground state is a magnetically and electronically inhomogeneous state characterized by FM clusters (on a broad spectrum of length scales) embedded in a non-FM matrix. The clusters have a mean correlation length of 50 Å at 4 K, although magnetic inhomogeneity occurs across a broad spectrum of length scales, evidenced by a highly inhomogeneous ground state. This magnetically inhomogeneous state manifests an intercluster magnetoresistance effect, which is a phenomenon of importance in understanding the CMR effect in perovskite cobaltites as well as for manganites. In terms of electronic properties, generally the perovskite cobaltites have insulating/semiconducting ground states. He et al. [24] have reported finite spin states, formation of ferromagnetic droplets (of nanometers in size), and a magneto-electronically phase-separated state characterized by FM clusters embedded in a non-FM matrix over a doping range $0.04 < x < 0.22$ in $La_{1-x}Sr_xCoO_{3-\delta}$. For $x > 0.22$ on the other hand, behaves as a uniform FM metal [24].

2.3.1 Electronic Phase Separation in $La_{0.7-x}Ln_xCa_{0.3}CoO_3$ (Ln = Pr, Nd, Gd, and Dy)

Figure 2.24 shows the result of magnetic measurements for the $La_{0.7-x}Ln_xCa_{0.3}CoO_3$ perovskites with (Ln = La, Pr, Nd), which signifies how the magnetic transition is sensitive to the substitution of the smaller cations in place of rare earth La^{3+}. $La_{0.7}Ca_{0.3}CoO_{3-}$ exhibits a ferromagnetic transition ($T_C \sim 175$ K), $Pr_{0.7}Ca_{0.3}CoO_3$ and $Nd_{0.7}Ca_{0.3}CoO_3$ do not show distinct ferromagnetic transitions [23].

There is a slight increase in the susceptibility of $Pr_{0.7}Ca_{0.3}CoO_3$ around 75 K, but this is not due to a genuine ferromagnetic transition. On the basis of the $\langle r_A \rangle$ values, the ferromagnetic T_C's of $Pr_{0.7}Ca_{0.3}CoO_3$ and $Nd_{0.7}Ca_{0.3}CoO_3$ expected to

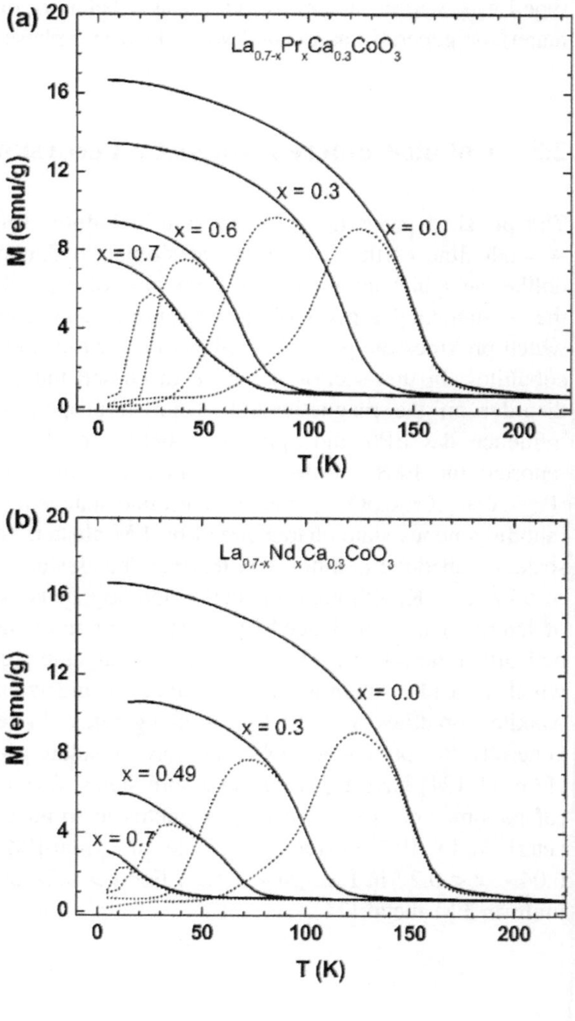

Fig. 2.24 Temperature dependence of magnetization for $La_{0.7-x}Ln_xCa_{0.3}CoO_3$ with **a** Ln = Pr and **b** Ln = Nd. ZFC data in broken curves and FC data in solid curves (at 1 kOe) (adapted from Ref. [23])

Fig. 2.25 Variation of the ferromagnetic T_C with x in $La_{0.7-x}Ln_xCa_{0.3}CoO_3$. The inset shows the variation of T_C with $\langle r_A \rangle$ (Å) (adapted from Ref. [23])

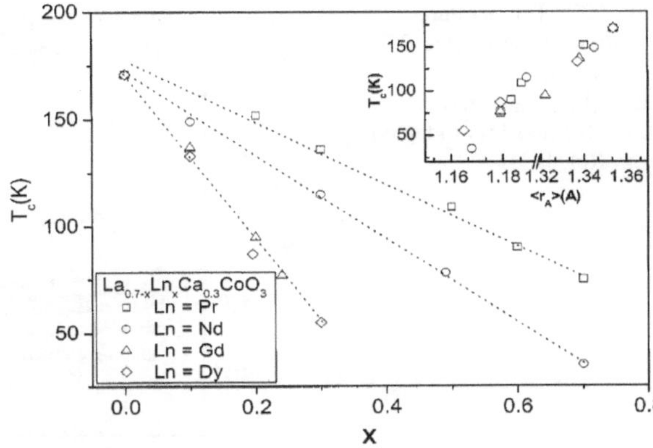

be well above 100 K. The FM T_C values in the four series of perovskite cobaltites are plotted against x in Fig. 2.25. The T_C value decreases linearly with increasing x.

The M(H) hysteresis curves do not show saturation in all the compositions. The absence of saturation is a characteristic of a glassy system. Furthermore, the remanent magnetization, M_r, decreases with increase in x or decrease in $\langle r_A \rangle$. The magnetization, M, and M_r increase with $\langle r_A \rangle$, but their values become rather low when $\langle r_A \rangle \leq 1.18$ Å (Fig. 2.26).

The electrical resistivities of the perovskite cobaltites show similar trends to the magnetic properties. Figure 2.27 shows resistivity for two series of perovskite

Fig. 2.26 Variation of **a** the magnetic moment, μ_B, **b** remanent magnetization, M_r, and **c** the electrical resistivity in $La_{0.7-x}Ln_xCa_{0.3}CoO_3$ with $\langle r_A \rangle$ (Å) (50 K) (adapted from Ref. [23])

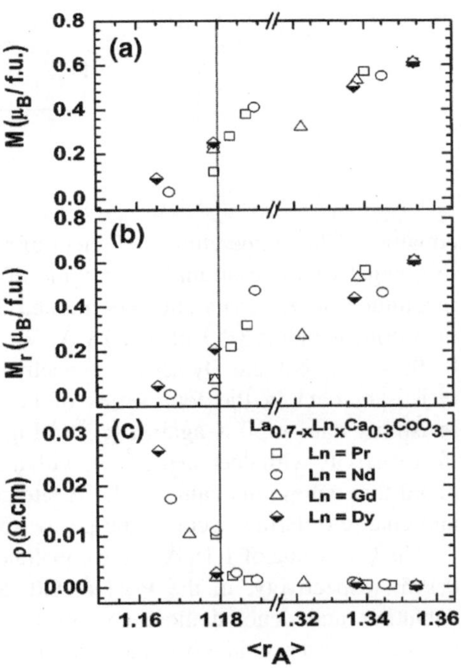

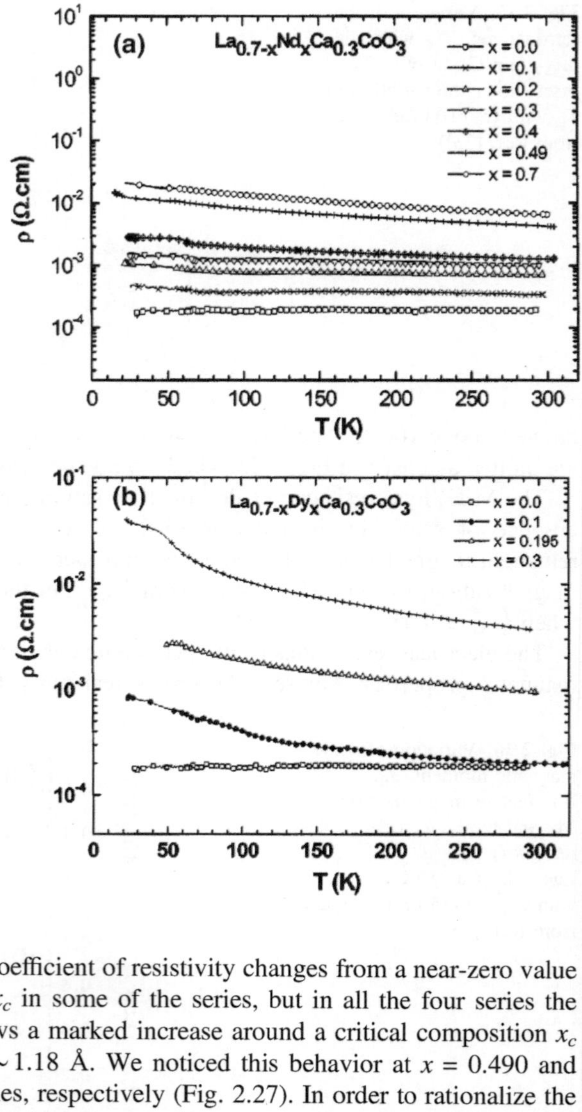

Fig. 2.27 Temperature variation of the electrical resistivity of $La_{0.7-x}Ln_xCa_{0.3}CoO_3$ (Ln = Nd and Dy) (adapted from Ref. [23])

cobaltites. The temperature coefficient of resistivity changes from a near-zero value to a negative value around x_c in some of the series, but in all the four series the magnitude of resistivity shows a marked increase around a critical composition x_c or a critical radius $\langle r_A^c \rangle$ of ~1.18 Å. We noticed this behavior at $x = 0.490$ and 0.195 for the Nd and Dy series, respectively (Fig. 2.27). In order to rationalize the resistivity data in the four series of perovskite cobaltites, we have plotted the resistivity data at 50 K against $\langle r_A \rangle$ in Fig. 2.26c. There is a noticeable increase in the resistivity with decreasing $\langle r_A \rangle$, with a change in slope around 1.18 Å. It is to be noted that below this value of the A-site cation radius, electronic phase separation and charge ordering occur in the perovskite manganites [10].

The $\langle r_A \rangle$ value of 1.18 Å in the cobaltates corresponds to $x \approx 0.6$, 0.49, 0.24, and 0.195, respectively, in the Pr, Nd, Gd, and Dy substituted series of perovskite cobaltites and denoted these compositions as x_c [23]. It appears that the small magnetic cluster regime becomes prominent around x_c or $\langle r_A^c \rangle$. The low-temperature peak in the AC susceptibility data is frequency dependent; the small magnetic cluster regime at small $\langle r_A \rangle$ in the $La_{0.7-x}Ln_xCa_{0.3}CoO_3$ system is considered as the regime to be magnetically inhomogeneous. The inhomogeneous nature of the perovskite cobaltites prevails over the entire range of compositions ($x = 0.0$–0.7). Hence, with a composition close to x_c or $\langle r_A \rangle$, there is a marked change in the distribution of the magnetic species. Thus, when $x < x_c$ or $\langle r_A \rangle > \langle r_A^c \rangle$, relatively large ferromagnetic clusters or domains are present in the system, resulting in large magnetization and T_C values. When $x > x_c$ or $\langle r_A \rangle < \langle r_A^c \rangle$, the magnetic clusters become small in size. The ferromagnetic clusters being hole-rich, the electrical resistivity data show changes around the same compositions as the magnetization data, the compositions with $\langle r_A \rangle > \langle r_A^c \rangle$ exhibiting lower resistivities and near-zero temperature coefficients of resistivity. Owing to the change in the nature of magnetic species around x_c or $\langle r_A^c \rangle$ as a mere change in size distribution, we designated this as a case of phase separation since more than one transition is observed in the AC susceptibility data. The phase-separated regime here involves the coexistence of large ferromagnetic clusters, which are hole-rich and small clusters which are hole-poor [23].

Accordingly, AC susceptibility (Fig. 2.28a) of perovskite $La_{0.7}Ca_{0.3}CoO_3$ ($x = 0.0$, $\langle r_A \rangle = 1.354$ Å) shows frequency-independent transition around 150 K corresponding to the FM transition, while $Pr_{0.7}Ca_{0.3}CoO_3$ ($x = 0.7$, $\langle r_A \rangle = 1.179$ Å)

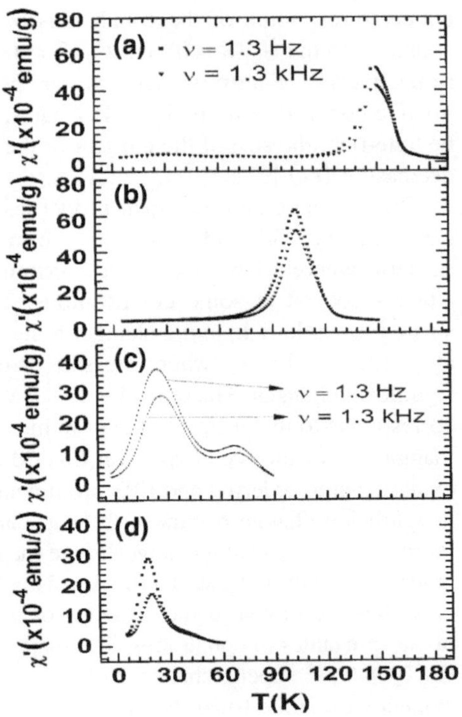

Fig. 2.28 Temperature variation of the AC susceptibility of **a** $La_{0.7}Ca_{0.3}CoO_3$, **b** $La_{0.4}Pr_{0.3}Ca_{0.3}CoO_3$, **c** $Pr_{0.7}Ca_{0.3}CoO_3$, and **d** $Nd_{0.7}Ca_{0.3}CoO_3$ at two different frequencies (adapted from Ref. [23])

shows two distinct transitions, the low temperature one with a greater frequency dependence (Fig. 2.28c). Perovskite cobaltite $La_{0.4}Pr_{0.3}Ca_{0.3}CoO_3$ ($\langle r_A \rangle = 1.194$ Å) also shows a single transition corresponding to the T_C which is frequency independent similar to $La_{0.7}Ca_{0.3}CoO_3$. The frequency-independent high-temperature transition in perovskite $Pr_{0.7}Ca_{0.3}CoO_3$ is due to the large magnetic clusters (akin to cluster glass [25]) as in $x = 0.0$ composition and the low-temperature transition is due to small magnetic clusters which seem to show some spin glass characteristics. Thus, with the increase in x or decrease in $\langle r_A \rangle$ for the $La_{0.7-x}Ln_xCa_{0.3}CoO_3$, the large ferromagnetic clusters seem to progressively give way to the small clusters, giving rise to magnetic phase separation.

The presence of very weak features at low temperatures in the AC susceptibility data of $La_{0.7}Ca_{0.3}CoO_3$ indicates that the proportion of small clusters is negligible. Whereas for $Nd_{0.7}Ca_{0.3}CoO_3$ a frequency-dependent low-temperature transition (around 20 K) due to the small magnetic clusters (Fig. 2.28d). It is instructive to discuss the nature of the spin states of cobalt in $La_{0.7-x}Ln_xCa_{0.3}CoO_3$ series. The magnetic moment of the cobalt ion provides an average magnetic moment value of 4.5 μ_B per cobalt ion in all the series of perovskite cobaltites. This value suggests that the cobalt ions are in the intermediate-spin (IS) and/or high-spin (HS) states. The IS and HS states of Co^{3+} correspond to the electronic configurations $t_{2g}^5e_g^1$ (S = 1) and $t_{2g}^4e_g^2$ (S = 2), respectively, and those of Co^{4+} to $t_{2g}^4e_g^1$ (S = 3/2) and $t_{2g}^3e_g^2$ (S = 5/2). The spin state transitions in the perovskite cobaltites have shown that at high temperatures, the cobalt ions are mostly in the IS or the HS state ref. At low temperatures, some of the cobalt ions may go to the low-spin (LS) state, corresponding to the t_{2g}^6 (S = 0) and t_{2g}^5 (S = 1/2) configurations in Co^{3+} and Co^{4+} ions, respectively. The ferromagnetic clusters present prominently at $x < x_c$ or $\langle r_A \rangle > \langle r_A^c \rangle$ involve cobalt ions in the IS or HS states. The ferromagnetic regime will therefore be hole-rich, the size of the clusters or the domains decreasing with increasing x or decreasing $\langle r_A \rangle$ [23].

The magnetic and electrical properties of polycrystalline $La_{0.7-x}Ln_xCa_{0.3}CoO_3$ (Ln = Pr, Nd, Gd, and Dy) series are understood in terms of a phase separation scenario wherein large carrier-rich ferromagnetic clusters and carrier-poor smaller clusters coexist at some compositions. The ferromagnetic clusters prominent at small x are hole-rich, and a change in the electrical resistivity behavior is observed at a critical value x_c, where the size distribution of magnetic clusters undergoes significant changes. The critical value of x in the four series of perovskite cobaltites corresponds to the critical value of radius, $\langle r_A^c \rangle$, of 1.18 Å, a value where rare earth manganites of the type $La_{0.7-x}Ln_xCa_{0.3}MnO_3$ (Ln = Nd, Gd, and Y) are known to exhibit charge ordering and EPS prominently [10]. It appears that around $\langle r_A^c \rangle$ or x_c, a significant change occurs in the e.g. bandwidth and the charge carriers become more localized, causing changes in the magnetic and electron transport properties. It is well to recall that the electrical resistivity and ferromagnetism in the perovskite cobaltites are linked to the presence of the Co^{3+}–O–Co^{4+} species with the appropriate spin states of cobalt ions. The magnetism in the perovskite cobaltites is due to Co^{3+}–O–Co^{4+} superexchange, but most of the $Ln_{1-x}A_xCoO_3$ also seems to show evidence for some frustration at low temperature, as though there is no long-range ferromagnetism. In order to understand the nature of these materials, we have further discussed the magnetic properties of $Ln_{0.7}Ca_{0.3}CoO_3$ and $La_{1-x}Sr_xCoO_3$ cobaltites, down to low temperatures.

2.3.2 *Spin Glass Behavior in $Ln_{0.7}Ca_{0.3}CoO_3$ (Ln = La, Pr, and Nd)*

Preliminary DC magnetic susceptibilities of polycrystalline as well single crystalline samples of $Ln_{0.7}Ca_{0.3}CoO_3$ series (Ln = La, Pr, and Nd) showed that while $La_{0.7}Ca_{0.3}CoO_{3-}$ clearly exhibits a ferromagnetic-type transition (T_C ~ 175 K), $Pr_{0.7}Ca_{0.3}CoO_3$ and $Nd_{0.7}Ca_{0.3}CoO_3$ do not show distinct ferromagnetic transitions. There is a slight increase in the magnetization of $Pr_{0.7}Ca_{0.3}CoO_3$ around 75 K, but not due to a long-range ferromagnetic transition. The magnetic behavior of a single crystal of $Pr_{0.7}Ca_{0.3}CoO_3$ is similar to that of the polycrystalline sample [23]. The large drop in the magnetic moment at low temperatures in the Pr and Nd derivatives is noteworthy. In order to understand the nature of these materials, we have discussed the magnetic properties of $Ln_{0.7}Ca_{0.3}CoO_3$ in detail, down to low temperatures. The electronic phase separation and associated magnetic properties of $Pr_{0.7}Ca_{0.3}CoO_3$ and $Nd_{0.7}Ca_{0.3}CoO_3$ arise because of the small average size of the A-site cations. In these perovskite cobaltites, the average radius (for orthorhombic structure) is less than 1.18 Å, which is the critical value only above which long-range ferromagnetism manifests [23]. It is known that increase in size disorder and decrease in size of the A-site cations favor EPS.

Figure 2.29 shows temperature dependence of the ZFC and FC magnetization of single crystalline perovskites $Ln_{0.7}Ca_{0.3}CoO_{-3}$ (Ln = La, Pr, or Nd) measured parallel and perpendicular to the c-axis in an applied field of 20 Oe.

Perovskite $La_{0.7}Ca_{0.3}CoO_3$ shows a distinct FM-type transition in the FC data around 170 K (T_C), while the ZFC data show a cluster glass transition around 95 K. Whereas, $Pr_{0.7}Ca_{0.3}CoO_3$ and $Nd_{0.7}Ca_{0.3}CoO_3$ cobaltites do not, however, show such a ferromagnet-type behavior. The irreversible temperature, T_{irr}, between the ZFC and FC in perovskites $Pr_{0.7}Ca_{0.3}CoO_3$ and $Nd_{0.7}Ca_{0.3}CoO_3$ persists up to 200 K unlike in the $La_{0.7}Ca_{0.3}CoO_3$. The T_{irr}, however, decreases with the increasing magnetic field. The inverse susceptibility data could be fitted to Curie–Weiss behavior with the extrapolated Weiss temperatures, θ_p, of 150 K, −180 K, and −340 K for the La, Pr, and Nd derivatives, respectively. The negative θ_p values in the latter cobaltites imply the presence of antiferromagnetic interactions in the high-temperature region, while for La derivative the interaction is FM.

The M(H) loops for $Pr_{0.7}Ca_{0.3}CoO_3$ and $Nd_{0.7}Ca_{0.3}CoO_3$ perovskites at different temperatures are presented in Fig. 2.30. These perovskite cobaltites show hysteresis loops at low temperatures (≤10 K) and a nonsaturating behavior up to 5 Tesla. The coercive field and remanent magnetization are almost the same in both the orientations for all three cobaltites.

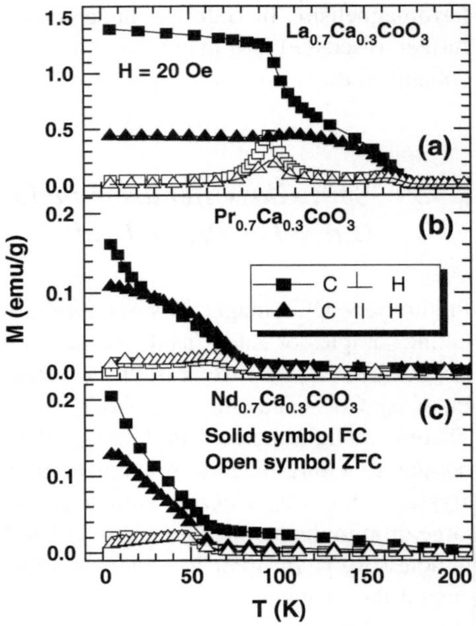

Fig. 2.29 Temperature dependence of the ZFC and FC magnetization, M, of $Ln_{0.7}Ca_{0.3}CoO_3$ where **a** Ln = La, **b** Ln = Pr, and **c** Ln = Nd, at H = 20 Oe measured parallel (*triangle*) and perpendicular (*square*) to the c-axis (adapted from Ref. [23])

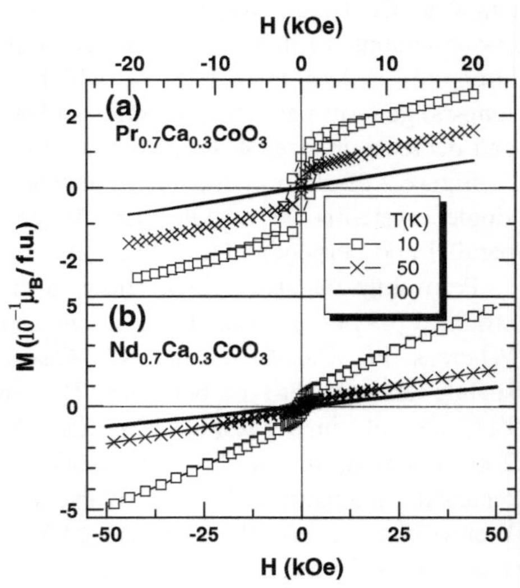

Fig. 2.30 The hysteresis cures for **a** $Pr_{0.7}Ca_{0.3}CoO_3$ **b** $Nd_{0.7}Ca_{0.3}CoO_3$ at different temperatures measured parallel to the c-axis (adapted from Ref. [23])

The magnetic behaviors of $Ln_{0.7}Ca_{0.3}CoO_{-3}$ perovskites have been discussed in terms of AC susceptibility and magnetic relaxation measurements which are useful to investigate the magnetic glassy behavior [23]. Figure 2.31 shows the AC susceptibility of $La_{0.7}Ca_{0.3}CoO_3$ below Curie temperature (T_C), which is in accordance with the low field ZFC magnetization. Two distinct peaks: a frequency-independent high-temperature peak (170 K) that indicates a ferromagnetic ordering and a low

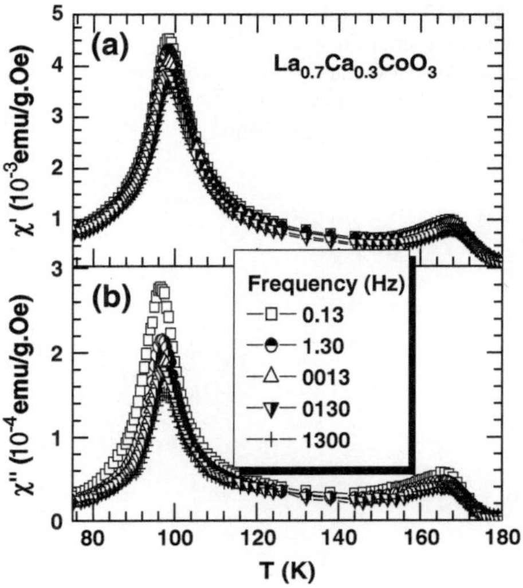

Fig. 2.31 The temperature dependence of the **a** In-phase and **b** Out-of-phase AC susceptibility of $La_{0.7}Ca_{0.3}CoO_3$ at different frequencies (adapted from Ref. [23])

temperature frequency-dependent peak at 100 K. The AC susceptibility for $Pr_{0.7}Ca_{0.3}CoO_3$ and $Nd_{0.7}Ca_{0.3}CoO_3$ exhibits a frequency-dependent maximum around 70 K for $Pr_{0.7}Ca_{0.3}CoO_3$ (Fig. 2.32). With increasing frequency, the peak value shifts toward higher temperatures. $Nd_{0.7}Ca_{0.3}CoO_3$ shows a peak around 55 K as shown in Fig. 2.32. Thus, with decrease in the average radius of the A-site cations, $\langle r_A \rangle$, the magnetic transition temperature as revealed by the AC susceptibility maximum shifts to lower temperatures.

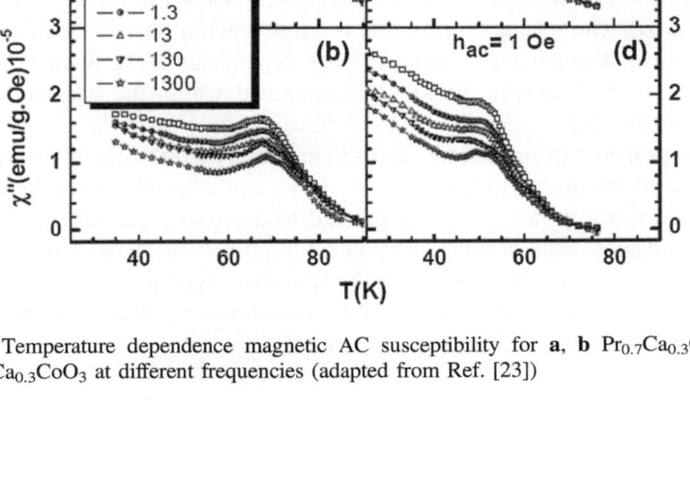

Fig. 2.32 Temperature dependence magnetic AC susceptibility for **a**, **b** $Pr_{0.7}Ca_{0.3}CoO_3$ and **c**, **d** $Nd_{0.7}Ca_{0.3}CoO_3$ at different frequencies (adapted from Ref. [23])

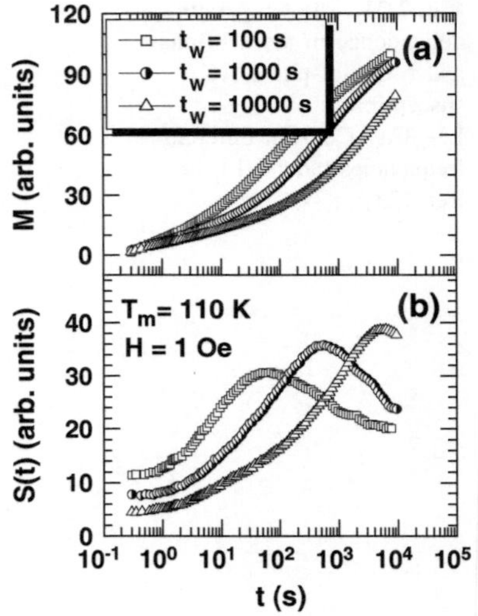

Fig. 2.33 ZFC-relaxation measurements on $La_{0.7}Ca_{0.3}CoO_3$ at T_m = 110 K for different waiting times (adapted from Ref. [23])

The ZFC time-dependent magnetic relaxation for $La_{0.7}Ca_{0.3}CoO_3$ is a consequence of nonequilibrium spin-glass-like state (Fig. 2.33). The long-time relaxations of the magnetization and aging phenomena well known in spin glasses [21] are commonly found in many other random magnetic systems. In Fig. 2.33 an aging behavior, similar to spin glasses revealed by an inflection point in the magnetization versus log (t) plot and a corresponding maximum in the relaxation rate curves at an observation time close to the wait time. The aging-dominated relaxation is strikingly similar to the behavior of conventional spin glasses. In the latter situation, aging is interpreted within the droplet model (or domain growth) for spin glasses to reflect the growth of equilibrium spin glass domains, with the maximum in the relaxation rate being associated with a crossover between quasi-equilibrium (from processes within ordered spin glass domains) and nonequilibrium dynamics (processes governed by effects at domain walls) [21].

Similarly, the time-dependent ZFC magnetization for $Pr_{0.7}Ca_{0.3}CoO_3$ and $Nd_{0.7}Ca_{0.3}CoO_3$ exhibits logarithmical dynamics below the transition temperature (70 and 55 K). The relaxation rate attains a maximum at the elapsed time, close to the wait time, indicating a pronounced age-dependent effect. Such a behavior is generally observed in spin glasses and explained within the droplet (or domain growth) model [21].

A key property to understand and model the dynamics of spin glasses is the occurrence of memory. A characteristic of the spin glass phase (ordinary or reentrant) is the memory behavior, whereas a disordered and frustrated ferromagnetic phase would show little or no memory effect. Figure 2.34a shows a memory dip for $La_{0.7}Ca_{0.3}CoO_3$ (at 85 K), indicating the spin-glass-type phase. At 110 K, on the other hand, a memory dip can barely be discerned, signifies that the system is confined in a disordered ferromagnetic phase.

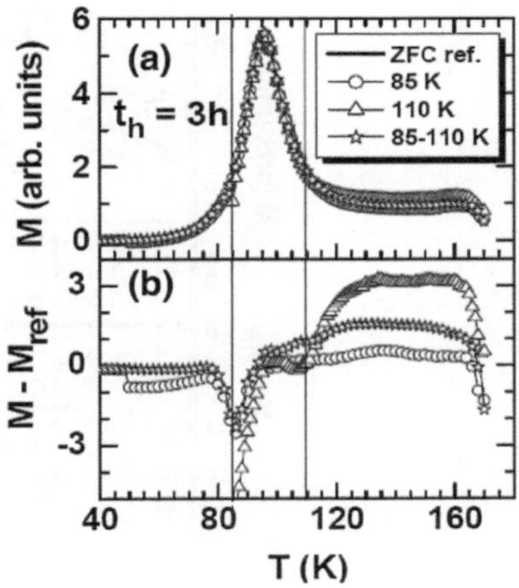

Fig. 2.34 The ZFC magnetization memory experiment on $La_{0.7}Ca_{0.3}CoO_3$; **a** the temperature dependence of ZFC magnetization, *M*, (*reference curve*) and memories of two temperature stops (at 85 and 110 K) during cooling each for 3 h and **b** the difference ($M–M_{ref}$) plot of the respective curves (adapted from Ref. [23])

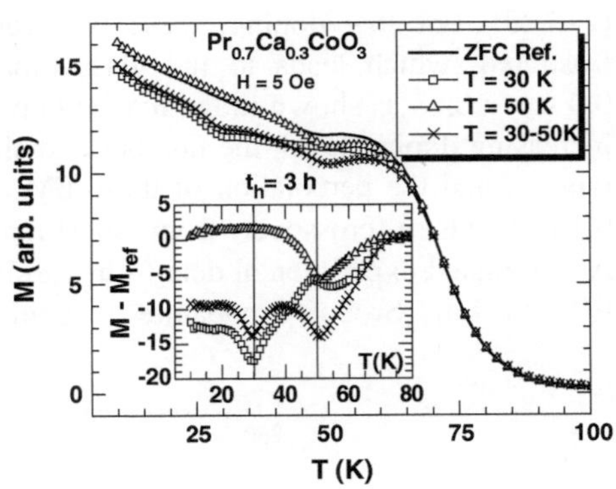

Fig. 2.35 The ZFC magnetization memory experiment on $Pr_{0.7}Ca_{0.3}CoO_3$ (adapted from Ref. [23])

Likewise, a memory effect is observed for $Pr_{0.7}Ca_{0.3}CoO_3$ (Fig. 2.35) and $Nd_{0.7}Ca_{0.3}CoO_3$ at different temperatures (Fig. 2.36). The memory dips are more prominent at low temperature for both the perovskites and correlate the spin-glass-like phases at the low temperatures [23].

2.3.3 Spin Glass Behavior in $La_{1-x}Sr_xCoO_3$

The hole-doped perovskite cobaltite $La_{1-x}Sr_xCoO_3$ is a model system to discuss the EPS and spin glass phenomenon because of the absence of charge ordering, FM insulating, and the long-range AFM ordering, which make it easier to probe and

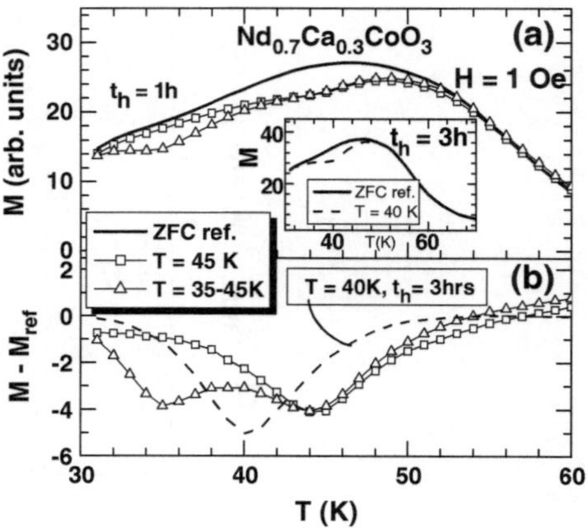

Fig. 2.36 The ZFC magnetization memory experiment on $Nd_{0.7}Ca_{0.3}CoO_3$ (adapted from Ref. [23])

understand the spin glass phenomenon [26–28]. Substituting divalent Sr^{2+} ions at trivalent La^{3+} sites in $LaCoO_3$ causes spontaneous nanoscopic phase separation where nanosized (i.e., 1–3 nm) hole-rich FM metallic clusters are embedded in a hole-poor insulating non-FM matrix [28]. The interaction between Co^{4+} and Co^{3+} is FM double exchange, whereas the Co^{3+}-Co^{3+} and Co^{4+}–Co^{4+} are antiferromagnetic [27, 28]. For low doping, these two competitive interactions are random and frustrated, which leads to the glassy magnetic behavior for the doping range $0.0 < x < 0.18$ as shown in the magnetic phase diagram of Fig. 2.37 [27, 28]. With increasing doping level, the number and size of the hole-rich FM clusters increase rapidly, and the percolation of these FM clusters at a critical doping value, x_p of 0.18, yields a crossover from short-range to long-range FM ordering [28]. Accumulated experimental data from various high-resolution probes reveal that the EPS in $La_{1-x}Sr_xCoO_3$ cobaltite is confined to a well-defined doping range,

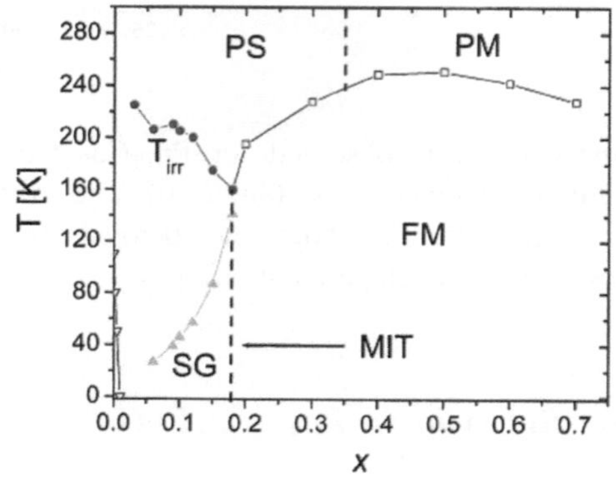

Fig. 2.37 Magnetic phase diagram for $La_{1-x}Sr_xCoO_3$. *PS* paramagnetic semiconductor, *PM* paramagnetic metal, *FM* ferromagnetic metal, *SG* spin glass, *MIT* metal-insulator transition, and T_{irr} is the irreversibility temperature which marks the bifurcation of ZFC and FC dc magnetization curves (adapted from Ref. [28])

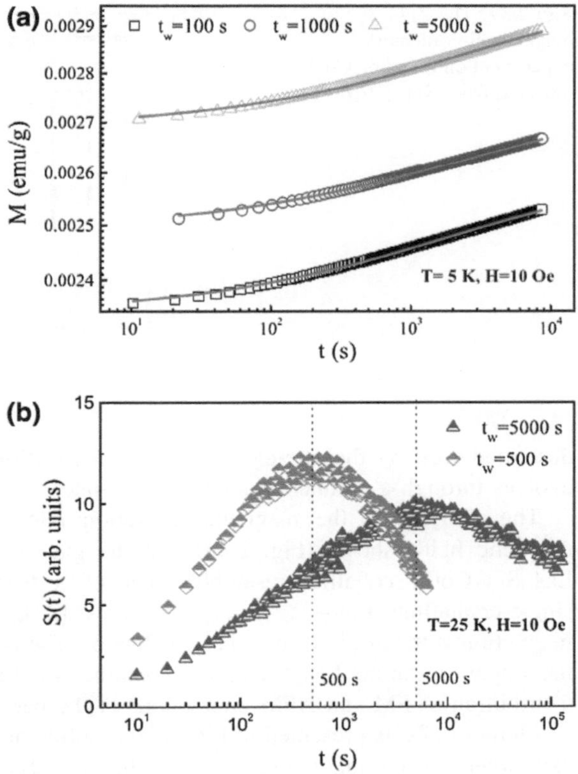

Fig. 2.38 ZFC-relaxation measurements on $La_{0.9}Sr_{0.1}CoO_3$ (adapted from Ref. [29])

$0.04 < x < 0.22$, which covers both the FM and glassy magnetic states (Fig. 2.37). It is fascinating to explore whether the phase-separated states in $La_{1-x}Sr_xCoO_3$ cobaltites exhibiting glassy magnetic behavior resemble that of a canonical/atomic spin glass or a superparamagnetic system or an assembly of strongly interacting magnetic clusters. Wu et al. [28] have pointed out that at lower Sr doping ($x < 0.18$) the system enters a mixed phase that displays the characteristics of both a spin glass and a ferromagnet. A cusp in the zero-field-cooled DC magnetization, a frequency-dependent peak in the AC susceptibility and time-dependent effects in both DC and AC magnetic properties all point toward glassy behavior. However, for high Sr doping ($x > 0.2$) the system exhibits unconventional ferromagnetism with a Curie temperature up to 250 K, which is interpreted in terms of the coalescence of short-range-ordered FM clusters.

Earlier studies on polycrystalline $La_{1-x}Sr_xCoO_3$ ($0 < x < 0.15$) cobaltites showed (Fig. 2.37) the presence of multiple glassy magnetic phases at low temperatures and also are superparamagnetic below the irreversibility temperature T_{irr} [27, 28]. Recent, investigations by Khan et al. [29], in single crystalline $La_{0.9}Sr_{0.1}CoO_3$ cobaltite of magnetic relaxation and memory effects below the freezing temperature T_f, reveal the characteristics of the spin glass phase. The magnetic relaxation is

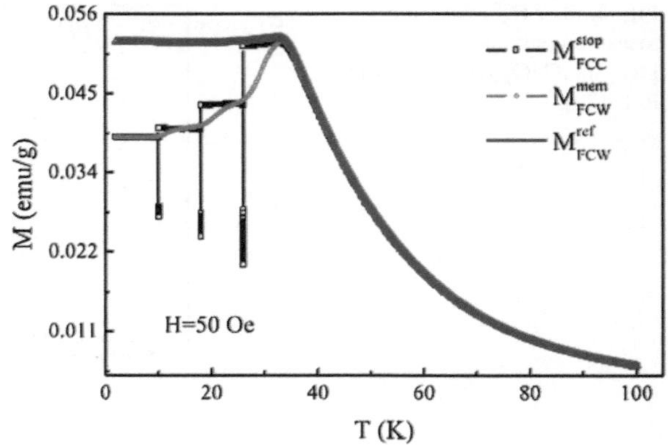

Fig. 2.39 The ZFC magnetization memory experiment on $La_{0.9}Sr_{0.1}CoO_3$ (adapted from Ref. [29])

described well by the stretched exponential function and shows that the system evolves through a number of intermediate states.

The analysis of the magnetic relaxation rate at different temperatures and magnetic fields shows (Fig. 2.38) that the glassy behavior of single-crystalline $La_{0.9}Sr_{0.1}CoO_3$ cobaltite resembles that of a true spin glass phase akin to single-crystalline $Ln_{0.7}Ca_{0.3}CoO_3$ cobaltites [23], where only the intercluster interaction is the origin of the glassy behavior. The observed spin glass behavior in the single-crystalline $La_{0.9}Sr_{0.1}CoO_3$ cobaltite is believed to be due to the random distribution of FM and AFM interactions in the perovskite [29].

Memory effects presented in Fig. 2.39 at different temperature and field-cycling experiments, show that the single-crystalline $La_{0.9}Sr_{0.1}CoO_3$ cobaltite is capable of retaining the magnetization history even for a large change in the magnetization. The presence of memory dips in the ZFC magnetization suggests that glassy magnetic behavior in single-crystalline $La_{0.9}Sr_{0.1}CoO_3$ cobaltite originates from spin–spin interaction, unlike the independent relaxation of clusters, which gives rise to superparamagnetic behavior. The effects of positive and negative temperature changes on the reversion of the original spin configuration suggest that the memory phenomena in this perovskite cobaltite follow the hierarchical model of spin glass [21, 22].

References

1. E.O. Wollan, W.C. Koehler, Phys. Rev. **100**, 545 (1955); W.C. Koehler, E.O. Wollan, J. Phys. Chem. Solids **2**, 100, (1957); C.N.R. Rao, A.K. Cheetham, R. Mahesh, Chem. Mater. **8**, 2421 (1996); E.L. Nagaev, Phys. Usp **39**, 781 (1996); C.N.R. Rao, Chem. Eur. J. **2**, 1499 (1996); A. P. Ramirez, J. Phys.: Condens. Matter **9**, 8171 (1997); C.N.R. Rao, B. Raveau (eds.), *Colossal Magnetoresistance, Charge-Ordering and Related Properties of Manganese Oxides* (World Scientific, Singapore, 1998); A. Mareo, S. Yunoki, E. Dagotto, Science **283**, 2034 (1999); C. N.R. Rao, A. Arulraj, A.K. Cheetham, B. Raveau, J. Phys.: Condens. Matter **12**, R83 (2000); Y. Tokura, *Colossal Magnetoresistive Oxides* (Gordon and Breach, NewYork, 2000)

2. E. Dagotto, T. Hotta, A. Moreo, Phys. Rep. **344**, 1 (2001); E. Dagotto, *Nanoscale Phase Separation and Colossal Magnetoresistance: The Physics of Manganites and Related Compounds* (Springer, Berlin, New York, 2003); N.D. Mathur et al., Phys. Today **56**, 25, (2003); C.N.R. Rao, A.K. Kundu, M.M. Seikh, L. Sudheendra, Dalton Trans. **19**, 3003 (2004); V.B. Shenoy et al., Phys. Rev. Lett., **98**, 097201 (2007); V.B. Shenoy, C.N.R. Rao, Phil. Trans. R. Soc. A **366**, 63 (2008); V.B. Shenoy et al., Chem. Phys. Chem. **7**, 2053, (2010); B. Raveau, M.M. Seikh, *Cobalt Oxides: From Crystal Chemistry to Physics* (Wiley-VCH, 2012)
3. L.M. Rodriguez-Martinez, J.P. Attfield, Phys. Rev. B **54**, R15622 (1996); J.P. Attfield, Chem. Mater. **10**, 3239 (1998)
4. J.M.D. Teresa, M.R. Ibarra, P.A. Algarabel, C. Ritter, C. Marquina, J. Blasco, J. Garcia, A.D. Moral, Z. Arnold, Nature **386**, 256, (1997); D.E. Cox, P.G. Radealli, M. Marezio, S-W. Cheong, Phys. Rev. B **57**, 3305 (1998); S. Marcone et al., Phys. Rev. B **68**, 094422 (2003)
5. P.G. Radealli, R.M. Ibberson, D.N. Argyriou, H. Casalta, K.H. Andersen, S.-W. Cheong, J.F. Mitchell, Phys. Rev. B **63**, 172419 (2001)
6. G.S. Bhalla, S. Selcuk, T. Dhakal, A. Biswas, A.F. Hebard, Phys. Rev. Lett. **102**, 077205 (2009); Y. Murakami et al., Nat. Nanotech. **5**, 37 (2010); T.Z. Ward, Z. Gai, H.W. Guo, L.F. Yin, J. Shen, Phys. Rev. B **83**, 125125 (2011); D. Niebieskikwiat, R.D. Sánchez, J. Phys.: Condens. Mater. **24**, 436001 (2012)
7. A. Moreo, S. Yunoki, E. Dagotto, Science **283**, 2034 (1999); M. Fäth et al., Science **285**, 1540, (1999); M.R. Freeman et. al. Science **294**, 1484, (2001); J. C. Loudon et al., Nature **420**, 797, (2002); Ch. Renner, G. Aeppli, B.G. Kim, Y.A. Soh, S.W. Cheong, Nature **416**, 518 (2002); L. Zhang et. al., Science **298**, 805, (2002); L. Ghivelder, F. Parisi, Phys. Rev. B **71**, 184425 (2005); E. Dagotto, Science, **309**, 257 (2005); V. B. Shenoy et al., Phys. Rev. Lett. **98**, 097201 (2007); J. Tao et al., Phys. Rev. Lett. **103**, 097202 (2009); Liang et al., Nanoscale Res. Lett. **9**, 325 (2014) and references therein
8. C.N.R. Rao, A.R. Raju, V. Ponnambalam, S. Parashar, N. Kumar, Phys. Rev. B **61**, 594 (2000); S. Srivastava, N.K. Pandey, P. Padhan, R.C. Budhani, Phys. Rev. B **62**, 13868 (2001); R.C. Budhani, N.K. Pandey, P. Padhan, S. Srivastava, R.P.S. M. Lobo, Phys. Rev. B **65**, 14429 (2002)
9. M. Uehara, S. Mori, C.H. Chen, S-W. Cheong, Nature (London) **399**, 560 (1999); P.B. Littlewood, Nature (London) **399,** 529 (1999). V. Podzorov, M. Uehara, M.E. Gershenson, T. Y. Koo, S.W. Cheong, Phys. Rev. B **61**, R3784 (2000)
10. L. Sudheendra, C.N.R. Rao, J. Phys.: Condens. Mater. **15**, 3029 (2003); A.K. Kundu, P.V. Vanitha, C.N.R. Rao, Solid State Comm. **125,** 41 (2003)
11. A.M. Balagurov, VYu. Pomjakushin, D.V. Sheptyakov, V.L. Aksenov, P. Fischer, L. Keller, OYu. Gorbenko, A.R. Kaul, N.A. Babushkina, Phys. Rev. B **64**, 24420 (2001)
12. H. Terashita, J.J. Neumeier, Phys. Rev. B **63**, 174436 (2001)
13. J.M. de Teresa, M.R. Ibarra, J. Garcia, J. Blasco, C. Ritter, P.A. Algarabel, C. Marquina, A. del Moral, Phys. Rev. Lett. **76**, 3392 (1996)
14. I.G. Deac, S.V. Diaz, B.G. Kim, S.-W. Cheong, P. Schiffer, Phys. Rev. B **65**, 174426 (2002)
15. A.K. Kundu, M.M. Seikh, K. Ramesha, C.N.R. Rao, J. Phys.: Condens. Mater. **17**, 4171 (2005); A.K. Kundu, P. Nordblad, C.N.R. Rao, J. Phys.: Condens. Mater. **18**, 4809 (2006)
16. A. Maignan, C. Martin, M. Hervieu, B. Raveau, J. Hejtmanek, Solid State Commun. **107**, 363 (1998); I.O. Troyanchuk, D.D. Khalyavin, S.V. Trukhanov, H. Szymczak, J. Phys.: Condens. Matter **11**, 8707 (1999)
17. P.M. Woodward, D.E. Cox, T. Vogt, C.N.R. Rao, A.K. Cheetham, Chem. Mater. **11**, 3528 (1999)
18. C. Ritter, R. Mahendiran, M.R. Ibarra, L. Morellon, A. Maignan, B. Raveau, C.N.R. Rao, Phys. Rev. B **61**, R9229 (2000)
19. P.V. Vanitha, C.N.R. Rao, J. Phys.: Condens. Mater. **13**, 11707 (2001)
20. H. Kuwahara, Y. Tomioka, A. Asamitgu, Y. Moritomo, Y. Tokura, Science **270**, 961 (1995); P.V. Vanitha, R.S. Singh, S. Natarajan, C.N.R. Rao, J. Solid. State. Chem. **137**, 365 (1998); P.V. Vanitha, A. Arulraj, A.R. Raju, C.N.R. Rao, C.R. Acad. Sci. Paris, **2**, 595 (1999)

21. J.A. Mydosh, in Spin Glasses: An Experimental Introduction (Taylor and Francis, London, 1993); K. Binder, A.P. Young, Rev. Mod. Phys. 58, 801 (1986)
22. D.N.H. Nam, R. Mathieu, P. Nordblad, N.V. Khiem, N.X. Phuc, Phys. Rev. B **62**, 1027 (2000)
23. A.K. Kundu, K. Ramesha, R. Seshadri, C.N.R. Rao, J. Phys.: Condens. Mater. **16**, 7955 (2004); A.K. Kundu, P. Nordblad, C.N.R. Rao, Phys. Rev. B **72**, 144423 (2005); A.K. Kundu, P. Nordblad, C.N.R. Rao, J. Solid State Chem. **179**, 923 (2006)
24. C. He, S. El-Khatib, J. Wu, J.W. Lynn, H. Zheng, J.F. Mitchell, C. Leighton, Euro. Phys. Lett., **87**, 27006 (2009); F. Guillou, Q. Zhang, Z. Hu, C.Y. Kuo, Y.Y. Chin, H.J. Lin, C.T. Chen, A. Tanaka, L.H. Tjeng, V. Hardy, Phys. Rev. B **87**, 115114 (2013); D. Phelan, K.P. Bhatti, M. Taylor, S. Wang, C. Leighton, Phys. Rev. B **89**, 184427 (2014) and references therein
25. M. Itoh, I. Natori, S. Kubota, K. Matoya, J. Phys. Soc. Jpn. **63**, 1486 (1994)
26. J. Wu, J.W. Lynn, C.J. Glinka, J. Burley, H. Zheng, J.F. Mitchell, C. Leighton, Phys. Rev. Lett. **94**, 037201 (2005); D. Fuchs et al., Phys. Rev. B **71**, 92406 (2005); M.W. Haverkort et al., Phys. Rev. Lett. **97**, 176405 (2006); A.K. Kundu et al., J. Solid State Chem. **180**, 1318, (2007); J. Yu et al., Phys. Rev. B **80**, 052402, (2009); C. He et al., Phys. Rev. B **80**, 214411, (2009)
27. P.M. Raccah et al., Phys. Rev. **155**, 932 (1967); V.G. Bhide, D.S. Rajoria, C.N.R. Rao, G.R. Rao, V.G. Jadhao, Phys. Rev. B **12**, 2832 (1975); M.A. Senaris Rodriguez et al., J. Solid State Chem. **118**, 323 (1995); V. G. Sathe et al., J. Phys.: Condens. Mater. **8**, 3889 (1996); M. Imada et al., Rev. Mod. Phys. **70**, 1039, (1998); R. Caciuffo et al., Phys. Rev. B **59**, 1068 (1999); R. Ganguly et al., J. Phys. Condens. Matter **13**, 10911, (2001); V.G. Prokhorov et al, Phys. Rev. B **66**, 132410 (2002); R. Mahendiran et al., Phys. Rev. B **68**, 24427, (2003); L. Sudheendra et al., Ferroelectrics **306**, 227 (2004); A. Ghoshray et al., Phys. Rev. B **69**, 064424, (2004); M.J.R. Hoch et al., Phys. Rev. B **69**, 014425, (2004); S. Tsubouchi et al. Phys. Rev. B **69**, 144406, (2004); A.K. Kundu et al., Solid State Commun. **134**, 307, (2005); S.R. Giblin et al., Euro. Phys. Lett., **70**, 677 (2005)
28. J. Wu, C. Leighton, Phys. Rev. B **67**, 174408 (2003); P.L. Kuhns, M.J.R. Hoch, W.G. Moulton, A.P. Reyes, J. Wu, C. Leighton, Phys. Rev. Lett. **91**, 127202 (2003)
29. N. Khan, P. Mandal, D. Prabhakaran, Phys. Rev. B **90**, 024421 (2014)

Chapter 3
Ordered-Disordered Perovskite Cobaltites

3.1 Introduction

Last few decades, there has been extensive research on ABO_3-type perovskite oxides of the general formula $Ln_{1-x}A_xBO_3$ (Ln = trivalent lanthanide, A = divalent alkaline earth, B = transition metal) [1]. Particularly, the perovskite cobaltites were discovered in the 1950s and the magnetic ordering was first reported in 1960s [2]; since then there are lots of interesting phenomena reported in the literature. Some of the novel properties of the perovskite cobaltites are known for some time, especially the crystal structure transformation, the paramagnetic (PM) to ferromagnetic (FM) transition at Curie temperature (T_C) and the associated insulator–metal transition (T_{IM}) and so on. The discovery of colossal magnetoresistance (CMR) in doped manganites has renewed great interest in perovskite oxides since the early 1990s [1]. In 1997, large value of magnetoresistance was reported for ordered perovskite cobaltites, $LnBaCo_2O_{5.4}$ (Ln = Eu, Gd) known as layered 112-phases [3]. This leads to enhanced interest because of their potential applications in improving magnetic data storage, magnetic sensors, etc. Moreover, the perovskite cobaltites have also attracted attention because of their possible applications as oxidation catalysts, gas sensors, solid oxides fuel cells, and oxygen separation membranes due to their stability in a wide range of oxygen contents, high oxygen diffusivity, and electron conduction [4]. Consequently, superconductivity was discovered in hydrated sodium cobaltite phase in 2003, since then cobaltites have received even more attention [1, 5]. Aside from potential applications, the cobaltites exhibit rich phase diagram spanning a wide range of magnetic properties and phenomena like charge ordering, orbital ordering, spin/cluster-glass behavior, electronic phase separation, etc. [1, 6–11]. These phenomena represent a combined interaction between the spin, the lattice, the charge, and the orbital degrees of freedom, which will provide better understanding of strongly correlated electronic behavior. Such interactions are manifested in single crystal, polycrystalline samples as well as in thin films. The properties of perovskite cobaltites can be tuned either

A.K. Kundu, *Magnetic Perovskites*, Engineering Materials,
DOI 10.1007/978-81-322-2761-8_3

by external factors or by chemical means. In certain critical range of cation doping, at the A-site of ABO_3-perovskite, the rare earth cobaltites exhibit simultaneous occurrence of ferromagnetism and metallicity, along with CMR in the vicinity of T_C or T_{IM} [7].

Some of the perovskite-based cobaltites are known to exhibit electronic inhomogeneities arising from the existence of more than one phase in crystals of nominally monophasic composition. This is understood in terms of electronic phase separation described recently in the literature [1, 11, 12]. Such a phenomenon occurs because of the comparable free energies of the different phases. The phase-separated hole-rich and hole-poor regions give rise to anomalous properties such as weak FM moments in an antiferromagnetic regime.1,10 A variety of magnetic inhomogeneities manifest themselves in $Ln_{1-x}A_xCoO_3$ depending on the various factors, such as the average radius of the A-site cation and size-disorder as well as external factors, such as temperature, magnetic field, etc. [6–11]. In the last few years electronic phase separation in cobaltites has attracted considerable attention [1, 10, 11].

In this book, we discuss the A-site cationic ordering and disordering effects on magnetic and electron transport properties for perovskite cobaltites. Perovskite cobaltites have two possible forms of the A-site cations distribution depending on the type of cations or the synthesis procedures [9–12]. The first reported compounds on perovskite cobaltites are the A-site disordered structure [2, 13], which have been investigated for last few decades, and the other one is A-site ordered perovskites possessing a layered 112-type structure [3]. The latter one discovered few years back, consists of oxide layers [LnO]–[CoO_2]–[AO]–[CoO_2] alternating along the *c*-axis [1, 3, 9, 12, 14, 15]. The ordering of Ln^{3+} and A^{2+} ions is favorable if the size difference is large between the A-site cations, hence smaller size Ln^{3+} and bigger A^{2+} ions easily form a layered 112-structure and till date it is reported only for $A^{2+} = Ba^{2+}$ cation, i.e., $LnBaCo_2O_{5+\delta}$ ($0 \leq \delta \leq 1$) series [3, 9, 12, 14–16]. In contrast, as the size difference becomes smaller, for example between La^{3+} and Ba^{2+}, the disordered cubic perovskite becomes more stable and special treatment is required to obtain the 112 ordered structure, as reported for the order–disorder phenomena observed in the perovskites of the system La–Ba–Co–O [12, 16]. The CoO_2 layers in ordered cobaltites are free from the random potential which would otherwise arise from the Coulomb potential and/or local strain via the random distribution of A-site cations (Ln^{3+}/A^{2+}). Both ordered and disordered cobaltites exhibit similar features like ferromagnetism, insulator–metal transition, and magnetoresistance in a certain temperature range, yet prominent differences are evidenced in their properties. The ferromagnetism in these cobaltites is considered to be due to favorable Co^{3+}–O–Co^{4+} interactions. The effect of random distribution of A-site cations or disordered perovskite structure arising from the chemical disorder as observed in the conventional random doping at the A-site in ABO_3-perovskite. For rare earth manganites, Attfield et al. [17] have reported the role of A-site cationic radius, $\langle r_A \rangle$, and size disorder parameter, σ^2, on the T_C and/or T_{IM} with variation of these parameters. In the present article, we have also discussed the role of $\langle r_A \rangle$, and σ^2 for disordered cobaltites $Ln_{0.5}A_{0.5}CoO_{3+\delta}$ (Ln = La, Nd, Gd, and A = Ba, Sr) which unusually influence the magnetic and electronic properties [13, 18, 19]. These interesting phenomena are related to structural disordering caused by the substitution of A^{2+} ions in place of Ln^{3+} and we have briefly presented the effect of size disorder σ^2 for Ln = Nd, Gd [18, 19]. The structural disorder in cobaltites is 3D, while a layered 2D structure is adopted by the 112-phase ordered cobaltites, $LnBaCo_2O_{5+\delta}$. The crystallographic structures for perovskite cobaltites with an integral number of oxygen ions per formula unit are well known, no consensus has been reached for noninteger compounds. Basic knowledge of the crystallographic structure of a compound is of particular importance for determining magnetic structures, since these two properties are closely related. Hence, we have presented first the crystal structure description for both the (ordered-disordered) perovskite cobaltites before discussing the physical properties.

3.2 Crystal Structure of Perovskite Cobaltites

In general, the rare earth ordered-disordered cobaltites crystallize in the perovskites structure with various types of superstructures also evidenced. The disordered ABO_3-perovskite is a simple cubic structure (Pm-3 m) as in the previous section of this book. However, many perovskite deviate a little from this structure even at room temperature. The stability of the perovskite structure of cobaltites depends on the relative size of the Ln/A and Co ions in $Ln_{1-x}A_xCoO_3$. In disordered cobaltites Ln/A cation is surrounded by eight corner sharing CoO_6 octahedra, which build a 3D network. As a consequence, for the disordered perovskite cobaltites $Ln_{0.5}Ba_{0.5}CoO_3$, the crystal structure is reported as cubic or rhombohedral for Ln = La, whereas for other lanthanides, such as Pr, Nd, Gd, Dy the systems crystallize in the orthorhombic structure with different space groups [13, 18]. The structural Rietveld analysis for disordered $Ln_{0.5}Sr_{0.5}CoO_3$ perovskites reveal that the structure is rhombohedral for Ln = La, Pr Nd and that for Gd, the structure is orthorhombic as reported for Ba-doped compounds [13].

Similarly, the layered 112-type ordered cobaltites $LnBaCo_2O_{5+\delta}$ can be found in several crystallographic symmetries at room temperature. The variation of oxygen stoichiometry ($0 \leq \delta \leq 1$) in ordered 112-type cobaltites lead to various structures as well as different cobalt and oxygen coordination's, such as pyramidal, octahedral, and/or the mixing of both environments for the Co-ions [14, 15]. In the following, the room temperature structures for different values of 'δ' will be discussed. The crystal structure of the stoichiometric $LnBaCo_2O_5$ (Ln = Eu, Gd, Tb, Ho) cobaltite is tetragonal with *P4/mmm* space group (unit cell $a_p \times a_p \times 2a_p$; where a_p is defined as pseudo cubic cell parameters) [14, 15]. This corresponds to a doubling of the original perovskite unit cell along the *c*-direction due to alternating BaO and LnO_δ layers (Fig. 3.1a). The layered structure is best observed in the most oxygen deficient case $LnBaCo_2O_5$, because it is assumed that the oxygen ions are absent only in the Ln-layer [14, 15]. For $\delta = 0$ the Co^{2+} and Co^{3+} ions (ratio 1:1) are all within square base pyramids formed by their five oxygen neighbors.

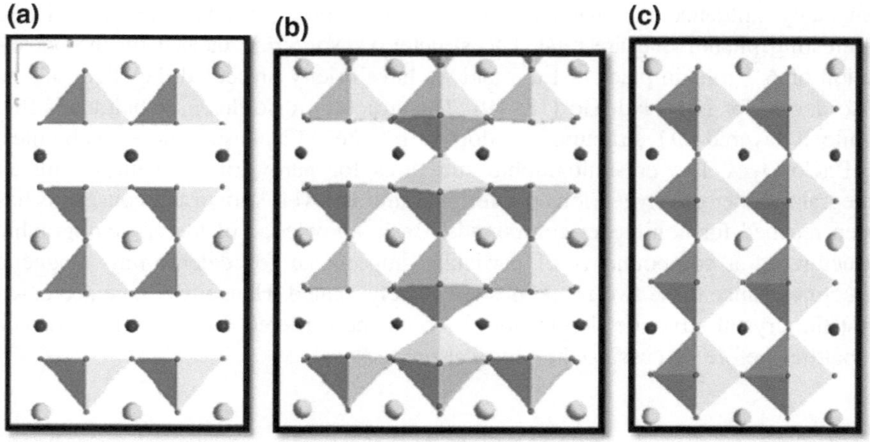

Fig. 3.1 Layered 112-type ordered $LnBaCo_2O_{5+\delta}$ ($0 \leq \delta \leq 1$) cobaltites with **a** $\delta = 0.0$; $LnBaCo_2O_5$, **b** $\delta = 0.5$; $LnBaCo_2O_{5.5}$ and **c** $\delta = 1.0$; $LnBaCo_2O_6$. All structures show a 1:1 ordering of the Ba (*green spheres*) and Ln (*blue spheres*) layers along *c*-axis

For nonstoichiometric cobaltites $LnBaCo_2O_{5+\delta}$ ($0 < \delta < 1$) the crystal structure is more complex: due to oxygen vacancy ordered super-structures can arise, which vary with oxygen content. The oxygen content is strongly dependent on the size of the lanthanides [14]. Clearly, a strong correlation exists between the Ln^{3+} radius and the amount of oxygen the compound can accommodate. The first structural study of an ordered oxygen deficient perovskite of type $LnBaCo_2O_{5+\delta}$ was presented for the series Ln = Pr to Ho [14]. Although, the series has been investigated in details for almost all lanthanide elements of the periodic table, interestingly there was no such report on the first member of this series, i.e., for $LaBaCo_2O_{5.5}$. Recently, Rautama et al. [16] have reported $LaBaCo_2O_{5.5}$ and characterized by neutron diffraction, electron microscopy and magnetic studies, showing that at room temperature the structure is 112-layered orthorhombic $a_p \times 2a_p \times 2a_p$ supercell with *Pmmm* space group (Fig. 3.1b). In contrast, for Ln = Pr–Ho, all of the x-ray diffraction patterns could be indexed using a tetragonal structure $a_p \times a_p \times 2a_p$ with *P4/mmm* symmetry [14]. However, electron diffraction measurements additionally revealed two kinds of superstructures, depending on the radius of the lanthanide (or the oxygen content). For larger lanthanides (Pr, Nd, Sm, Eu, Gd, and Tb) a doubling of one lattice parameter is observed, corresponding to an orthorhombic *Pmmm* $a_p \times 2a_p \times 2a_p$ supercell, whereas for smaller lanthanides (Ho and Dy) a tripling of two lattice parameters is noticed, as in a $3a_p \times 3a_p \times 2a_p$ supercell. It was suggested [14] that ordering of oxygen vacancies is at the origin of the observed superstructures. Importantly, the authors found that the oxygen vacancies are located uniquely in the LnO_δ layers (apical positions). By changing the oxygen content the superstructures either vanish or change from one to the other: Reducing oxygen content in $GdBaCo_2O_{5+\delta}$ from $\delta = 0.4$ to 0 destroys the superstructure. On the other hand in $HoBaCo_2O_{5.3}$ an increase of δ from 0.3 to 0.4 leads to a change in superstructure from $3a_p \times 3a_p \times 2a_p$ to $a_p \times 2a_p \times 2a_p$. High resolution electron microscopy supported the assumption that the ordering of the oxygen vacancies is responsible for the superstructures. Burley et al. and Pralong et al. [14] have reported $NdBaCo_2O_{5+\delta}$ for various oxygen stoichiometries. The $\delta = 0$ compound has a tetragonal structure with $a_p \times a_p \times 2a_p$ supercell where the Nd layer incorporates all the oxygen vacancies. For the slightly higher oxygen content $\delta = 0.38$ the structure is derived from the $\delta = 0$ one, but with oxygen ions inserted randomly into the NdO_δ layers. Oxygen vacancy ordering in *b*-direction resulting in an orthorhombic $a_p \times 2a_p \times 2a_p$ supercell unit cell with *Pmmm* symmetry is reported for $\delta = 0.5$, in agreement with other reported structure [14]. The oxidized material with $\delta = 0.69$ is again described by a tetragonal $a_p \times a_p \times 2a_p$ supercell unit cell, but a very weak peak originating from a doubling of the unit cell along the *b*-direction was detected.

Finally, for stoichiometric cobaltites $LnBaCo_2O_6$ all Co-ions (Co^{3+} and Co^{4+}; ratio 1:1) are in octahedral environment. In the $LnBaCo_2O_6$ series, the first member Ln = La could be synthesized in the two forms, ordered and disordered as shown by neutron diffraction and electron microscopy [9, 12]. At room temperature and in normal synthesis condition the crystal structure is cubic for the disordered cobaltite, which means the La^{3+} and Ba^{2+} ions are distributed randomly on the *A*-site. This is due to the smaller size differences between the cations as discussed earlier. Nevertheless, the ordered 112-phases $LaBaCo_2O_{5.5}$ and $LaBaCo_2O_6$ are obtained for the same composition in special synthesis conditions [12, 16], which crystallize at room temperature in orthorhombic and tetragonal structures respectively. Therefore, a layered 112-type ordered perovskite structure $LnBaCo_2O_6$ with a smaller lanthanide also exist, again with a tetragonal unit cell $a_p \times a_p \times 2a_p$ with *P4/mmm* symmetry, but require special synthesis conditions as reported by Pralong et al. [14]. To conclude the ordered cobaltites phase, a summary of the most commonly used models for compounds with oxygen content $\delta = 0, 0.5, 6$ at room temperature is given (Fig. 3.1):

1. $LnBaCo_2O_5$ has the tetragonal structure with $a_p \times a_p \times 2a_p$ supercell (*P4/mmm* symmetry). Ln^{3+} layers alternate with BaO layers along the *c*-axis (oxygen vacancies in Ln layer).
2. $LnBaCo_2O_{5.5}$ has an orthorhombic structure with $a_p \times 2a_p \times 2a_p$ supercell (*Pmmm* symmetry). $LnO_{0.5}$ layers alternate with BaO layers along the *c*-axis. Along the *b*-direction the oxygen vacancies are ordered. This leads to an alternation of CoO_5 pyramids and CoO_6 octahedra along the *b*-direction.
3. $LnBaCo_2O_6$ is again tetragonal unit cell $a_p \times a_p \times 2a_p$ with *P4/mmm* symmetry. LnO layers alternate with BaO layers along the *c*-axis (no oxygen vacancies in LnO layer).

We will now discuss in brief the various parameters to obtain different phases (ordered-disordered) of La–Ba–Co–O cobaltites. The synthesis of $LaBaCo_2O_{5.5}$, maintaining the "$O_{5.5}$" stoichiometry and a perfect layered ordering of La^{3+} and Ba^{2+} cations, is delicate due to their small size difference which favors their statistical distribution. Moreover, the larger size of La^{3+} compared to other lanthanides allows large amounts of oxygen to be inserted, so that the disordered $La_{0.5}Ba_{0.5}CoO_3$ perovskite is more easily formed under normal conditions. Thus, the successful synthesis of ordered phase requires several steps, using soft chemistry method, and the strategy was to control the order-disorder phenomena in this system by means of two synthesis parameters, temperature, and oxygen partial pressure. In order to favor the ordering of the La^{3+} and Ba^{2+} cations, the synthesis temperature was as low as possible, and consequently a soft-chemistry synthesis route was used since it allows a high reactivity at low temperature. However, this condition is not sufficient alone to achieve a perfect ordering of these cations. The formation of La^{3+} layers is in fact favored by the intermediate creation of ordered oxygen vacancies, leading then to the 112-type layered nonstoichiometric cobaltites $LaBaCo_2O_{5+\delta}$, built up of layers of CoO_5 pyramids between which the La^{3+} smaller than Ba^{2+} cations can be interleaved. In this process, it is rather difficult to control the oxygen stoichiometry to "$O_{5.5}$." For this reason, the synthesis of the ordered $LaBaCo_2O_6$ phase was carried out initially, using high purity argon gas, followed by annealing in an oxygen atmosphere at specific temperature. Thereafter, the layered 112 cobaltite $LaBaCo_2O_{5.5}$ was obtained from the ordered $LaBaCo_2O_6$ phase by employing temperature controlled oxygen depletion method in inert atmosphere [16].

The X-ray powder diffraction of the three different phases exhibit nice crystallization with different structures. The disordered $La_{0.5}Ba_{0.5}CoO_3$ perovskite and ordered $LaBaCo_2O_6$ perovskite indexed with the cubic *Pm*-3 *m* and tetragonal *P4/mmm* structures, respectively [12]. The ordered perovskite shows a doubling of the cell parameter along the *c*-axis related to the 1:1 ordering of the LaO/BaO layers (Fig. 3.1c). The structure of ordered $LaBaCo_2O_{5.5}$ perovskite is very different from the ordered $LaBaCo_2O_6$ perovskite and indexed with the *Pmmm* orthorhombic structure [16]. Importantly, the La/Ba ordering in the ordered perovskite also involves a slight deformation of the perovskite sublattice with a dilatation of the a_p parameter within the LaO/BaO layers and a compression along the LaO/BaO layers stacking direction [12]. The disordered $La_{0.5}Ba_{0.5}CoO_3$ and ordered $LaBaCo_2O_6$ perovskite structures are also confirm by the transmission electron microscopy (TEM) investigations. The selected area electron diffraction (SAED) and the corresponding high resolution electron microscopy (HREM) images of the disordered perovskite (Fig. 3.2a) are indeed characteristic of a cubic perovskite. Whereas, for ordered $LaBaCo_2O_6$ perovskite the the SAED patterns corresponds to a tetragonal cell.

The doubling of one cell parameter with respect to the simple perovskite cell is noticed on the HREM image (Fig. 3.2b) and on the corresponding SAED [100] zone axis patterns (inset of Fig. 3.2b). Beside the SAED, the bright field images are also very important to obtain information about the size of the 112-type domains, in view of the possibility of twinning. For disordered cubic perovskite there is no twinning, as expected for cubic symmetry. However, in the case of 112-ordered perovskite*s* the crystal twinning is clearly evidenced (Fig. 3.3).

The structural refinements of ordered $LaBaCo_2O_{5.5}$ perovskite, shows a deviation from the ideal orthorhombic *Pmmm* structure, which is due to partial disorder

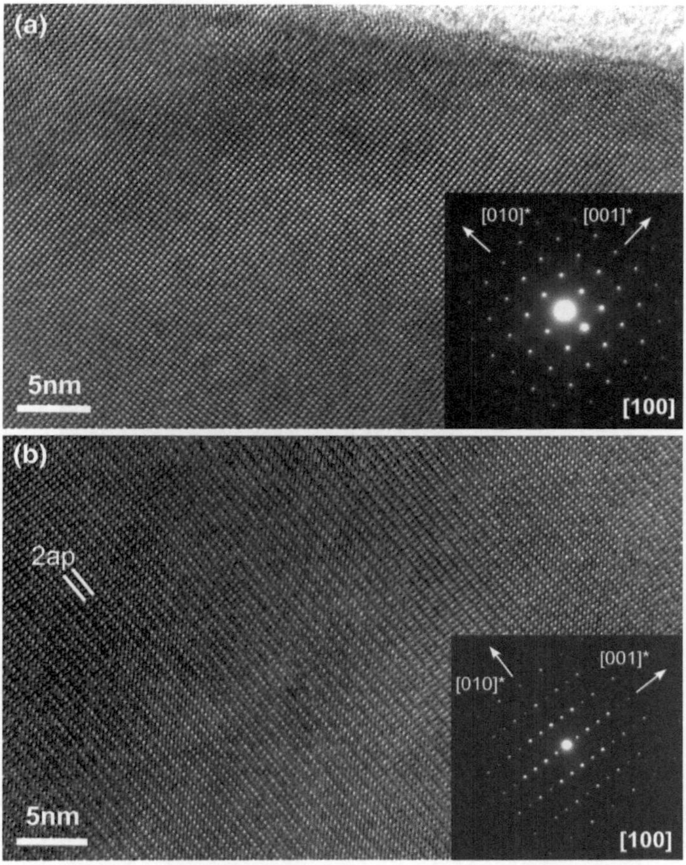

Fig. 3.2 The HREM images and the corresponding SAED patterns for **a** disordered $La_{0.5}Ba_{0.5}CoO_3$ and **b** ordered $LaBaCo_2O_6$ (adapted from Ref. [16])

of the oxygen atoms at the $LaO_{0.5}$ layers. The occupancy observed for the O6 and O7 sites by oxygen of about 90 and 10 %, respectively, instead of the expected 100 and 0 % [16]. The TEM images of ordered $LaBaCo_2O_{5.5}$ perovskite identified possible secondary phase, superstructures, and/or particular microstructural features. In Fig. 3.4, SAED pattern confirms that the observed spots can be indexed with the *Pmmm* orthorhombic $a_p \times 2a_p \times 2a_p$ structure. Additionally, two other features are also noticed: First, a twinned domains at a microscale level throughout the crystals, corresponding to the *Pmmm* $a_p \times 2a_p \times 2a_p$ structure (112-*type I*). Second, a very weak extra reflections that cannot be indexed considering only the *Pmmm* $a_p \times 2a_p \times 2a_p$ structure, in certain regions of the crystals. Considering, the literature on the 112-type compounds, a part of these extra spots was indexed considering the *Cmmm* $2a_p \times 4a_p \times a_p$ structure similar to $LaBaMn_2O_{5.5}$ perovskite [20]. This is in agreement with the NPD refinements where the occupancies for the

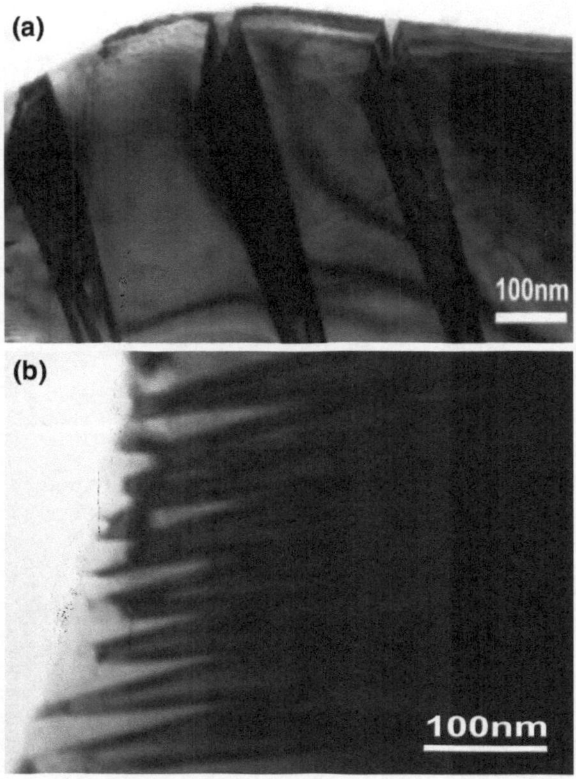

Fig. 3.3 Bright field images for ordered perovskite cobaltites **a** $LaBaCo_2O_{5.5}$ and **b** $LaBaCo_2O_6$ (adapted from Ref. [16])

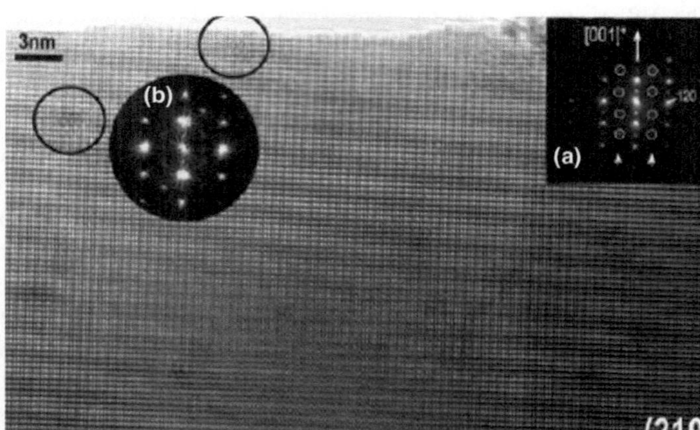

Fig. 3.4 The HREM images and the corresponding SAED patterns for ordered $LaBaCo_2O_{5.5}$ (adapted from Ref. [16])

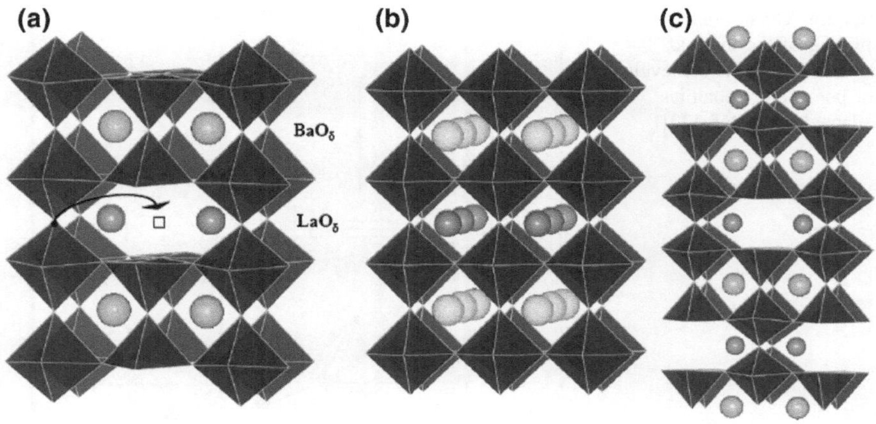

Fig. 3.5 Layered 112-type ordered $LaBaCo_2O_{5+\delta}$ cobaltites with **a** $\delta = 0.5$; $LnBaCo_2O_5$, (*type-I*) **b** $\delta = 1.0$; $LnBaCo_2O_6$ and **c** $\delta = 0.5$; $LnBaCo_2O_{5.5}$ (*type-II*) (adapted from Ref. [16])

O6 and O7 atomic positions could be locally attributed to a different vacancy/oxygen ordering (Fig. 3.5) leading notably to the existence of faulted zones having a centered structural motif (112-type II). This is also evidenced from SAED pattern and visible in HREM while observing through one of the directions (Fig. 3.4). In conclusion, the ordered $LnBaCo_2O_{5.5}$ perovskite consists mainly of 90 % 112-type I domains combined with 10 % of 112-type II (manganites) domains. It is also important to mentioned that there is no $3a_p \times 3a_p \times 2a_p$ superstructure for ordered $LaBaCo_2O_{5.5}$ perovskite, as evidenced for $HoBaCo_2O_{5.5}$ and $YBaCo_2O_{5.44}$ perovskites [14, 21].

3.3 Magnetic and Electron Transport Properties

The physical properties of doped rare earth perovskite cobaltites are influenced by two characteristic structural distortions. First, one is cooperative tilting of the CoO_6 octahedra which is essentially due to the doping effect. This distortion is a consequence of the mismatch of the ionic radius and various factors as discussed in the previous section. Second, the distortion arises from the Jahn-Teller (JT) effect due to Co^{3+} ion, which distorts the CoO_6 octahedra in such a way that there are long and short Co–O bonds (Fig. 3.6). This occurs below a characteristic temperature for particular compounds, as for instance at 180 K for the disordered $La_{0.5}Ba_{0.5}CoO_3$ perovskite [8]. It is well understood in terms of crystal field theory which describes how the *d*-electron of transition metal ions is perturbed by the chemical environment. The most effective distortion is the basal plane distortion (called Q_2 mode), with one diagonally opposite oxygen-pair displaced outwards and the other pair displaced inward. It is well established that a JT distortion involving a displacement

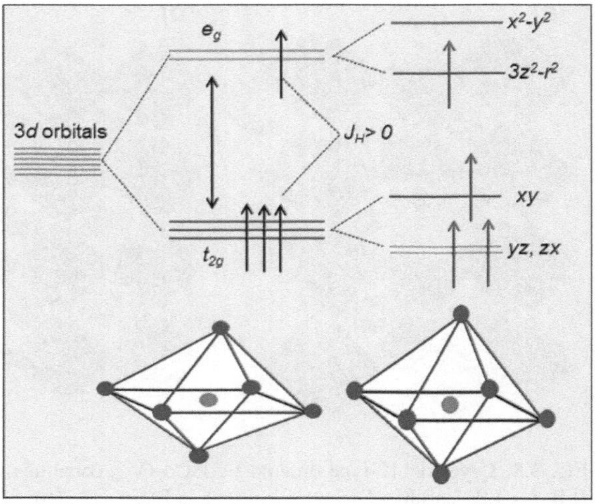

Fig. 3.6 Crystal-field splitting of the fivefold degenerate atomic 3d levels for perovskite cobaltites (adapted from Ref. [22])

of oxygen ions ≥0.1 Å can split the e.g.-band of the cobaltites (which forms the conduction band) and opens a gap at the Fermi level. The magnitude of the crystal field splitting of *d*-orbital determines whether the Co-ion occurs in the low-spin (LS), intermediate-spin (IS) or high-spin (HS) configuration. Figure 3.6, shows the schematic of $LaCoO_3$ band diagram to elucidate how the JT distortion splits the conduction band and makes the material insulating. The octahedral ligand environment around Co-ion splits the five *d*-orbitals into t_{2g}-triplet (d_{xy}, d_{yz} and d_{zx}) and e.g.-doublet (d^{22}_{x-y} and d^2_z) state.

In these type of perovskite cobaltites, the resulting crystal-field splitting, Δ_{cf}, between t_{2g} and e.g. orbital is around 2.06 eV as reported by Korotin et al. [22] for theoretical observation, although experimentally obtained values are around 1.2 and 0.9 eV [22]. Further splitting of the e.g. orbitals due to the JT effect opens a gap at the Fermi level. The intra-atomic exchange energy responsible for Hund's highest multiplicity rule, Δ_{ex} (or J_H), is smaller than Δ_{cf}, i.e., $\Delta_{ex} < \Delta_{cf}$ for Co^{3+} ion. Therefore, Co-ions are always in low spin state below 100 K for perovskite $LaCoO_3$ [22]. This perturbation induced electronic spin state transition in rare earth perovskite cobaltites has been of great interest in recent years [1]. The thermally driven spin state transition in cobaltites is a consequence of the subtle interplay between the crystal field splitting (Δ_{cf}) and the Hund's coupling energy (Δ_{ex}). The Δ_{cf} usually decreases as the temperature is increased, whereas Δ_{ex} is insensitive to temperature since it is an atomic quantity. The spin state of undoped $LaCoO_3$ (Co^{3+} ion) exhibits a gradual crossover with increasing temperature from the *LS* state ($t^6_{2g}e^0_g$; $S = 0$) to *IS* state ($t^5_{2g}e^1_g$; $S = 1$) at around 100 K and finally to *HS* state ($t^4_{2g}e^2_g$; $S = 2$) [22]. This results from the competition of the crystal field with energy Δ_{cf} (t_{2g}—e.g. splitting) and the interatomic (*Hund*) exchange energy Δ_{ex}, leading to redistribution of electrons between t_{2g} and e_g levels.

In perovskite $Ln_{1-x}A_xCoO_3$ cobaltites, the doping of A-site by A^{2+} cation induces the formal presence of Co^{4+}, which slightly reduces the JT distortion leading toward cubic structure. The JT distortion of CoO_6 octahedra has been reported by Fauth et al. [8] which is favored due to the *IS* state of Co^{3+} ($t_{2g}^5e_g^1$) and Co^{4+} ($t_{2g}^4e_g^1$). The transfer interaction of e_g-electrons is greater in rhombohedral or pseudocubic phase than in the orthorhombic phase because the ⟨Co–O–Co⟩ bond angle becomes closer to 180°. Clearly, the doping effect or the presence of Co^{4+} ion plays an important role in this material to provide the FM and metallic behavior by suppressing the JT distortion. The electron transport properties of cobaltites also depend on oxygen stoichiometry, and belong to the class either of metals or of semiconductors. Doped cobaltites are mixed valence materials, which mean that the Co-ions can carry different charges in the same site. The ratio between different Co-ion configurations is also determined by the oxygen stoichiometry as well as by the doping concentration, following the charge neutrality condition. Changes in the crystallographic and magnetic structures as well as the other physical properties can be induced by varying several parameters of either intrinsic hole/electron (doping, oxygen content) or extrinsic nature (temperature, pressure, magnetic field). Since only small changes of parameters (intrinsic/extrinsic) can cause a structural, magnetic, or electronic transition, and hence a variety of contradictory models or results exist for perovskite cobaltites [1]. Detailed discussions are therefore required to clarify the influence of each parameter on the physical properties of the perovskite cobaltites. In this article, special emphasis has been given on the cationic ordering and oxygen stoichiometry onto the crystallographic, magnetic, and electronic properties of few particular cobaltite systems. The magnetic structures, especially of the layered cobaltite systems with oxygen content of $O_{5.5}$ per unit formula, are still debated for some of the reported compounds. The determination of the magnetic structure is a complex task contrary to other materials, because the Co-ion in these materials can be in different spin states as discussed earlier.

3.3.1 Disordered Perovskite Cobaltites

Disordered rare earth cobaltites $Ln_{1-x}A_xCoO_3$ have been investigated for several years due to their novel magnetic and electronic properties which include temperature-induced spin state transitions, cluster glass-like behavior, electronic phase separation, magnetoresistance (MR) and so on [1, 6, 11, 13]. The physical properties of perovskite cobaltites are sensitively dependent on the doping concentration of the rare earth site. Doping brings up mixed valences in the Co-ions due to charge neutrality such as $(Ln^{3+}A^{2+})(Co^{3+}Co^{4+})O_3$. Therefore, substitution of Ln^{3+} by A^{2+} in $Ln_{1-x}A_xCoO_3$ will favor the transformation of Co^{3+} into Co^{4+} in same ratio of doping, as a result Co^{3+} and Co^{4+} will interact ferromagnetically obeying the Zener double-exchange (*DE*) mechanism [23]. The simultaneous observation of ferromagnetism and metallicity in cobaltites is explained by this mechanism, where the hopping of an electron from Co^{3+} to Co^{4+} via oxygen ion, i.e., where the Co^{3+} and Co^{4+} ions exchange takes place. The integral defining the exchange energy in such a system is nonvanishing only if the spins of the two *d*-orbitals are parallel. That is the lowest energy of the system is one with a parallel alignment of the spins on the Co^{3+} and Co^{4+} ions. Due to this, the spins of the incomplete *d*-orbitals of the adjacent Co-ion are accompanied by an increase in the rate of hopping of electrons and therefore by an increase in electrical conductivity. Thus, the mechanism which leads to enhanced electrical conductivity requires a FM coupling. On the other hand, Co^{3+}–Co^{3+} and Co^{4+}–Co^{4+} couple antiferromagnetically due to super-exchange interactions. Super-exchange interaction generally occurs between localized moments of ions in insulators. Goodenough et al. [23] pointed out that the FM interaction is governed not only by the *DE* interaction, but also by the nature of the super-exchange interactions. Whether the ferromagnetism in cobaltites (similar to manganites) is mediated by a DE mechanism or not is clearly not understood at present. However, the absence of half filled t_{2g} orbitals is providing core spin and strong Hund's rule coupling, unlike manganites, making this mechanism less feasible. It seems, the FM-metallic phase in cobaltites can be explained by the Zener-DE mechanism whereas super-exchange will fit for insulating state [23]. Hence, there will be always a competition between these two interactions to dominate one over another giving rise to a tendency of electronic phase separation in the system [1]. The growth of interest in perovskite cobaltites is due to the expectation that, in addition to the lattice, charge, and spin degrees of freedom found in many other transition metal oxides, the cobalt oxides also display a degree of freedom in the "spin-state" at the Co-site.

The physical properties of the cobaltites are sensitive to doping concentration at the rare earth site. Accordingly, the other parameters such as the average radius of the *A*-site cation, $\langle r_A \rangle$, and size-disorder parameter, σ^2, also vary due to doping at the perovskite *A*-site. These parameters crucially control the physical properties of disordered cobaltites. Disordered cobaltites of the type $Ln_{0.5}A_{0.5}CoO_3$, especially those with Ln = La, Nd, Gd and *A* = Ba, Sr are FM [13, 18, 19], many of them showing a metallic behavior as shown in Fig. 3.7. These properties arise because of the major influence of Co^{3+}–O–Co^{4+} interactions in these cobaltites. The FM T_C increases with the increase in the size of the *A*-site cations, $\langle r_A \rangle$. In the case of A = Ba, ferromagnetism occurs when Ln = La ($T_C \sim$ 190 K) and Nd ($T_C \sim$ 130 K), but for Ln = Gd, the material shows unusual magnetic behavior (Fig. 3.7c).

Furthermore, perovskite $Gd_{0.5}Ba_{0.5}CoO_3$ is an insulator and exhibits an electronic transition around 350 K with cationic ordering [13], whereas $La_{0.5}Ba_{0.5}CoO_3$ is metallic below FM T_C (Fig. 3.7d). Perovskite cobaltites $Gd_{0.5}Ba_{0.5}CoO_3$ which is charge-ordered at room temperature, shows a FM-like feature around 280 K, without reaching a saturation value of the magnetic moment or the highest moment achieved is rather low even in the higher applied field conditions [13, 18]. For $Gd_{0.5}Ba_{0.5}CoO_3$, the magnetic transition at 280 K has been defined as FM or meta-magnetic and for other perovskite cobaltites of this family, e.g., $La_{0.5}(Nd_{0.5})Ba_{0.5}CoO_3$ have also been designated as FM transitions. However, there are considerable differences amongst these perovskites. The magnetic transitions in perovskites $La_{0.5}(Nd_{0.5})Ba_{0.5}CoO_3$ are distinctly FM, showing a sharp increase in

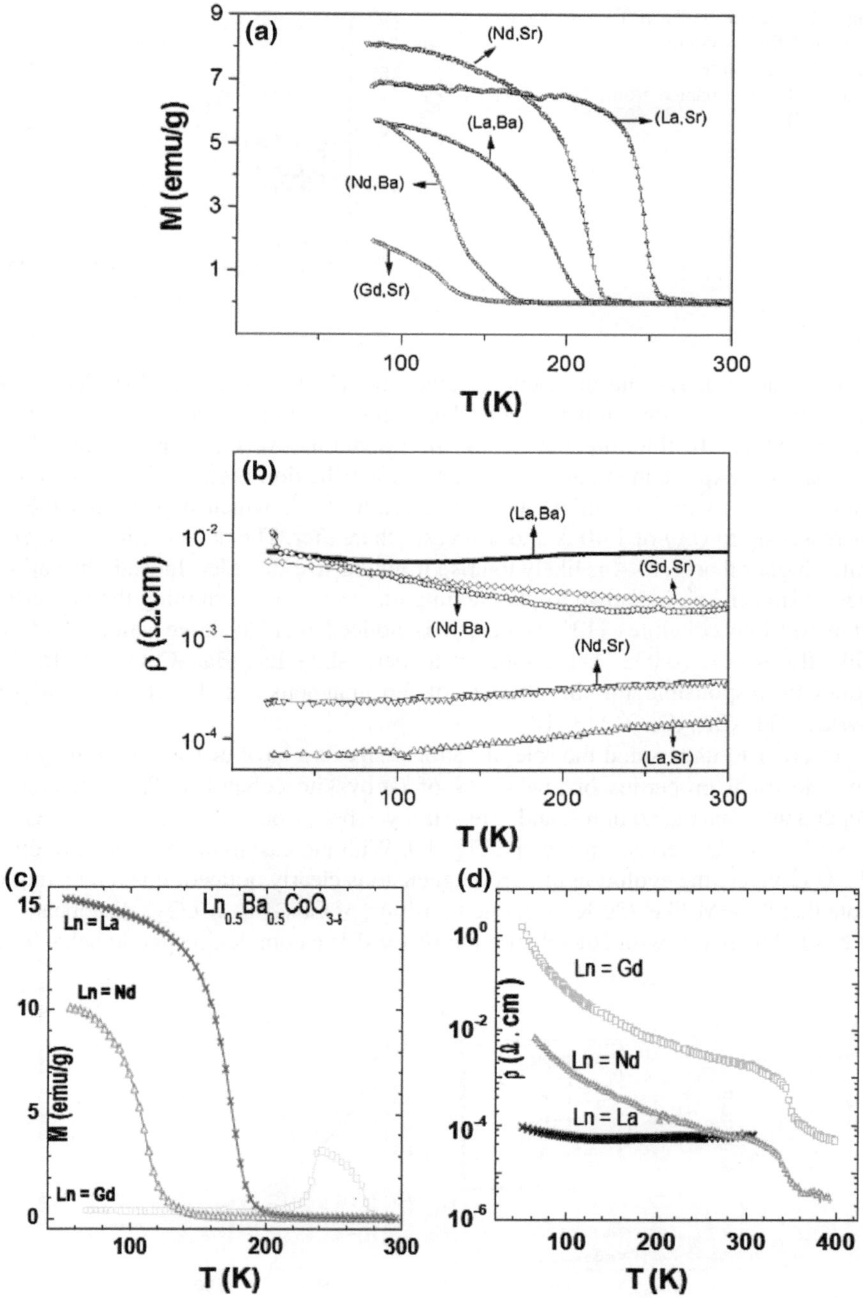

Fig. 3.7 Temperature dependent Magnetization and Resistivity for disordered perovskite cobaltites $Ln_{0.5}A_{0.5}CoO_3$ (adapted from Ref. [18])

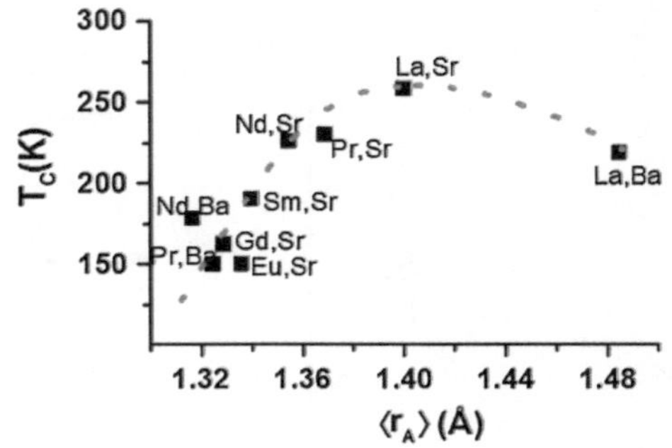

Fig. 3.8 Variation of FM T_C with $\langle r_A \rangle$ for disordered perovskite cobaltites $Ln_{0.5}A_{0.5}CoO_3$ (adapted from Ref. [13])

magnetization at T_C, and the value is rather low (130–190 K) [13, 18]. Perovskite $Gd_{0.5}Ba_{0.5}CoO_3$, with a much smaller lanthanide, could have been associated with lower FM T_C. In this direction a few series of $Ln_{0.5}A_{0.5}CoO_3$ perovskites have compared to explore the trends of magnetization behavior. Figure 3.8 shows the FM transitions T_C with the variation of ionic radius $\langle r_A \rangle$, which depicts that the T_C increases up to $\langle r_A \rangle$ of 1.40 Å and decreases thereafter. The decrease in T_C for $\langle r_A \rangle$ value higher than 1.40 Å is likely to arise from the size-disorder. Indeed, the cations size mismatch, σ^2, is known to play an important role in determining the properties of perovskite cobaltites [13]. Hence, it is noticed that the large value of σ^2 for $Gd_{0.5}Ba_{0.5}CoO_3$ (0.033 Å^2) compared to perovskite $La_{0.5}Ba_{0.5}CoO_3$ (0.016 Å^2) could be responsible for the absence of ferromagnetism and metallicity in perovskite $Gd_{0.5}Ba_{0.5}CoO_3$ [13, 18].

In order to understand the role of cationic size, we have discussed the magnetic and electrical properties of few series of perovskite cobaltites. The temperature dependent magnetization and resistivity behavior of perovskite $Gd_{0.5-x}Nd_xBa_{0.5}CoO_3$ series is presented Fig. 3.9. With increasing the Nd-substitution or the $\langle r_A \rangle$ value, the evolution of ferromagnetism is clearly noticed. It is interesting to note that the FM-like 280 K magnetic transition of $Gd_{0.5}Ba_{0.5}CoO_{2.9}$ disappears for $x \geq 0.1$. Moreover, with Nd-substitution of $x = 0.1$, a complex magnetic behavior is

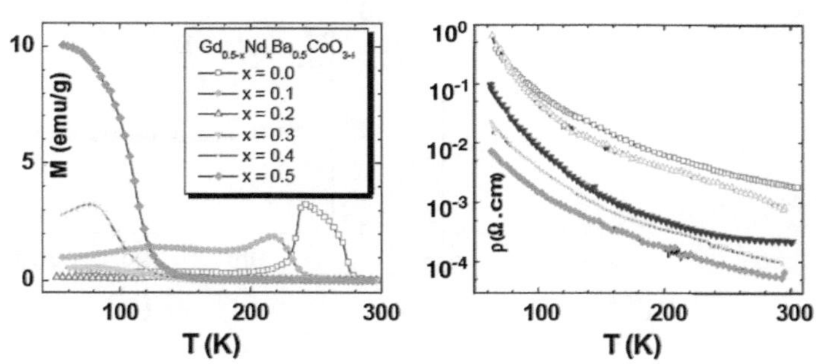

Fig. 3.9 Temperature dependent Magnetization and Resistivity for perovskites $Gd_{0.5-x}Nd_xBa_{0.5}CoO_3$ (adapted from Ref. [18])

appeared with a magnetic transition around 220 K. The $x = 0.3$ composition shows a weak magnetic transition around 125 K, and the transition is more prominent for $x = 0.4$. Similarly, for $Gd_{0.5-x}La_xBa_{0.5}CoO_3$ perovskites, there is no clear FM transition in the temperature range of 200–280 K for $0.1 < x < 0.25$. However, a distinct FM T_C apears at $x = 0.5$ in the case of Nd-perovskites, and for La-perovskites at $x = 0.4$. It is remarkable to notice that the FM transitions start emerging at low temperatures (<150 K) in these provskites with a $\langle r_A \rangle$, value close to 1.30 Å. This is explained by the fact that with increase in x, the FM clusters increase in size, which is eliminating the phase separation phenomena at lower value of x, caused by size disorder effect.

Likewise, it is noteworthy that in the perovskite cobaltites $Ln_{0.7}Ca_{0.3}CoO_3$ (Ln = La, *Pr*, Nd) the complex magnetism has been observed at low temperatures, the system approaching toward spin glass state for the higher size disorder [11]. Spin glass behavior is also reported for other perovskite $La_{1-x}Sr_xCoO_3$ ($x < 0.1$) and with increase in x ferromagnetism appears in the system akin to other perovskite cobaltites [13].

Perovskite cobaltites $Gd_{0.5-x}Nd_xBa_{0.5}CoO_{3-\delta}$ exhibit an insulating behavior throughout the temperature range, but the electrical resistivity decreases significantly with increase in x, particularly $x = 0.5$ composition exhibiting the lowest resistivity (Fig. 3.9). Similarly, for Ln = La, the resistivity value decreases with increasing x, becoming metallic for $x = 0.5$. It is observed that with increase in x, there is significant increase in $\langle r_A \rangle$ values in these provskite $Gd_{0.5-x}Ln_xBa_{0.5}CoO_3$ (Ln = La, Nd) series, this could be attributed to the effects of cation size which indirectly control the electronic band width as well as the energy band gap.

The role of size-disorder due to cation size mismatch, σ^2, could be more vividly understandable if we compare two series of perovskite cobaltites with fixed $\langle r_A \rangle$ values of 1.317 and 1.289 Å, corresponding to perovskites $Nd_{0.5}Ba_{0.5}CoO_3$ and $Gd_{0.5}Ba_{0.5}CoO_3$ respectively. The data, for a fixed $\langle r_A \rangle$ of 1.317 Å, show that the FM T_C decreases with increasing σ^2 (inset Fig. 3.10a), eventually destroying ferromagnetism at a high value of σ^2 (≈0.024 Å^2). Similarly, for a fixed $\langle r_A \rangle$ of 1.289 Å, are also confirming the cation size mismatch effect as shown in Fig. 3.11. It is observed that with decreasing σ^2, the magnetism of these perovskites changes significantly. Hence, for $\sigma^2 = 0.028$ Å^2, there is no such magnetic anomaly is observed similar to the perovskite $Gd_{0.5}Ba_{0.5}CoO_{2.9}$ ($T = 280$ K). However, with a lower value of $\sigma^2 = 0.021$ Å^2, a FM $T_C \sim 160$ K, and for $\sigma^2 = 0.018$ Å^2, the T_C value increased to 220 K. This is higher than the perovskite $La_{0.5}Ba_{0.5}CoO_3$. These phenomena signifies that the absence of long-range ferromagnetism in $Gd_{0.5}Ba_{0.5}CoO_3$, and the complex magnetism, like magnetic anomaly around 280 K, is entirely due to the disorder effect arising from the cation size mismatch. These size-disorder effect is at the origin of electronic phase separation as reported for other perovskite cobaltites [18]. The electrical resistivity of these series of cobaltites corroborates the results from the magnetic measurements. Figure 3.11b, exhibits the electrical resistivity for the series of the perovskites with increase in σ^2. Interestingly, disorder-induced insulator-metal transitions are noticed in both the series of perovskites and the systems with $\sigma^2 < 0.02$ Å^2 show metallic behavior.

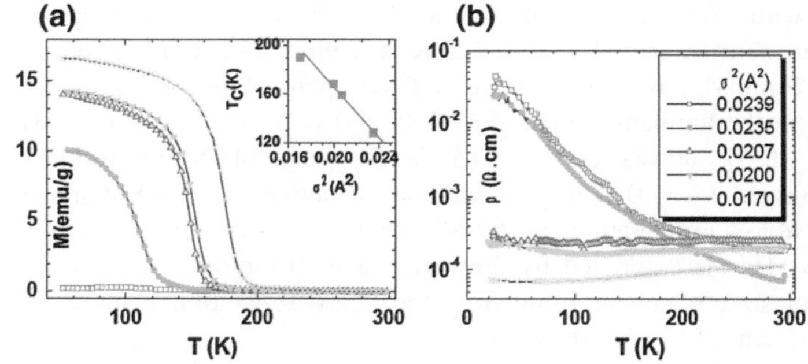

Fig. 3.10 Temperature dependent Magnetization and Resistivity for a fixed $\langle r_A \rangle$ of 1.317 Å, *inset* figure **a** shows the T_C–σ^2 plot (adapted from Ref. [18])

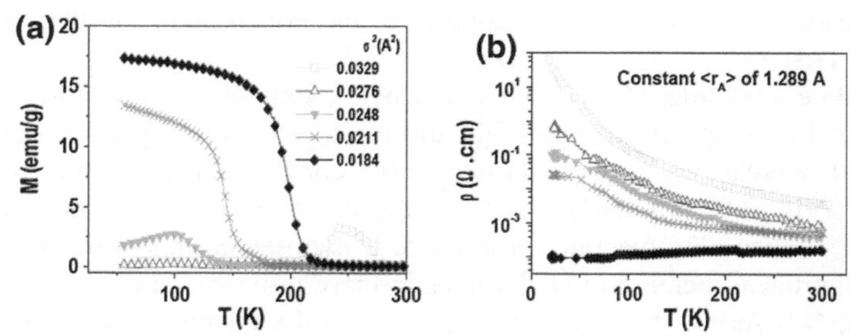

Fig. 3.11 Temperature dependent Magnetization and Resistivity for a fixed $\langle r_A \rangle$ of 1.289 Å (adapted from Ref. [18])

While disorder-induced meal-insulator transitions are common, size variance-induced insulator-metal transitions are indeed novel.

Further support of the previous discussion that cation size-disorder crucially determines the properties of perovskite cobaltite $Gd_{0.5}Ba_{0.5}CoO_{2.9}$ is provided in this section by the $Gd_{0.5}Ba_{0.5-x}Sr_xCoO_3$ series of perovskite cobaltites [18]. The $x = 0.5$ composition, corresponding to perovskite $Gd_{0.5}Sr_{0.5}CoO_3$, has a smaller $\langle r_A \rangle$ than $Gd_{0.5}Ba_{0.5}CoO_{2.9}$, and yet it shows FM transition. The 280 K magnetic anomaly of perovskite cobaltite $Gd_{0.5}Ba_{0.5}CoO_3$ disappears even when $x = 0.1$ and the apparent T_C increases with increasing x in this series. This behavior is clearly due to size disorder effect, since σ^2 decreases with increase in x. Accordingly, the present system exhibits an insulator-metal transition with increase in x or decrease in σ^2. It appears that a σ^2 value larger than 0.02 Å^2 generally destroys ferromagnetism in the cobaltites and changes the metal into an insulator as discussed earlier. In view of the interesting magnetic and electrical properties of the $Ln_{0.5}Ba_{0.5}CoO_3$ perovskite cobaltites, we have also presented a comparative study of Sr-doped cobaltites $Ln_{0.5}Sr_{0.5}CoO_3$ with Ln = La, Pr, Nd, Gd, and Dy. Magnetic properties of $La_{0.5}Sr_{0.5}CoO_3$ are well understood with FM T_C around 240 K. Also shows a frequency-dependent AC susceptibility maximum around 165 K, signifies a glassy ferromagnetism [19]. The perovskite cobaltites $La_{0.5}Sr_{0.5}CoO_3$, which was considered to be a good ferromagnetic metal a few years earlier, is actually a cluster-glass with some long-range magnetic ordering wherein frustration arise due to inter-cluster interactions at low temperatures [6]. The magnetic properties of these type of perovskites have been interpreted in terms of short-range magnetic ordering. Cluster glass behavior occurs at a high concentration of Sr ($x > 0.3$), where the coalescence of short-range FM clusters is proposed to occur [1, 6, 13]. ^{139}La NMR investigations confirm the coexistence of FM, paramagnetic, and cluster-glass type of phases in perovskite $La_{1-x}Sr_xCoO_3$ [1, 6, 13]. The phase separation in $La_{1-x}Sr_xCoO_3$ perovskite is linked to the formation of isolated nanoscopic FM clusters. Thus, the perovskites comprise FM clusters, paramagnetic matrices and spin-glass-like phases, are giving rise to complex glassy magnetism. More importantly, the glassy ferromagnetism is accompanied by electronic phase separation where the FM clusters exist within an antiferromagnetic (AFM) matrix. In this type of perovskite, the FM phase is metallic and the AFM phase is insulating in nature. Depending on x or the carrier concentration, we could have a complex system because the transition from the metallic to the insulating state is not prominent. Electronic phase separation with phases of different charge densities is generally give rise to nanometer scale clusters, which has been discussed earlier.

In the case of disordered perovskite $La_{1-x}Sr_xCoO_3$ the electronic phase separation is concomitant with the formation of isolated nanoscopic FM clusters. Evidence of phase separation in $La_{0.5}Sr_{0.5}CoO_3$ is experimentally obtained by the NMR studies [6]. The Mössbauer spectra reported by Bhide et al. [13] show the presence of a paramagnetic signal in addition to the six-finger pattern due to the FM-type species over a wide range of temperatures. The temperature dependent change in ferromagnetic-paramagnetic (FM/PM) ratio is reported for perovskite $La_{0.5}Sr_{0.5}CoO_3$ [19]. The FM/PM ratio increases with decreasing temperature, however the *PM* phase continues to exist well below T_C. This is vividly a direct evidence for electronic phase separation in perovskite $La_{0.5}Sr_{0.5}CoO_3$. It is of interesting to compare whether the glassy magnetism and phase separation phenomena exist in the other perovskites of the type $Ln_{0.5}Sr_{0.5}CoO_3$. The temperature-dependent magnetization data of perovskite $Gd_{0.5}Sr_{0.5}CoO_3$ exhibits behavior akin to magnetically disordered systems. What is particularly noteworthy is that the magnetization value below the FM T_C (110 K) in $Gd_{0.5}Sr_{0.5}CoO_3$ is much smaller than the other lanthanide perovskites. However, the carrier concentration or the Co^{3+}/Co^{4+} ratio remains constant in these perovskites, which could be due to the proportion of PM species relative to that of the FM species, those increases with decreasing the rare earth size. This is explained by the fact that the proportion of the magnetic clusters responsible for ferromagnetism decreases with decrease in $\langle r_A \rangle$.

In the disordered perovskite $Ln_{0.5}Sr_{0.5}CoO_3$ series, the FM T_C decreases with decreasing $\langle r_A \rangle$ (Fig. 3.12) which is a well-established behavior for these perovskite cobaltites [13, 19]. The magnetization value below FM T_C is associated with the proportion of the FM clusters, and hence we have compared in Fig. 3.12, the variation of magnetization with $\langle r_A \rangle$ for $Ln_{0.5}Sr_{0.5}CoO_3$ perovskite series as well as the size disorder effect. In order to correlate this disorder effect akin to

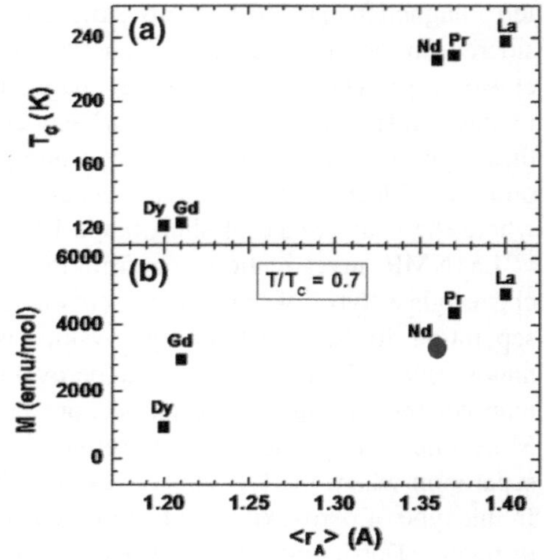

Fig. 3.12 Variation of **a** FM T_C and **b** Magnetization with $\langle r_A \rangle$ for perovskite cobaltites $Ln_{0.5}Sr_{0.5}CoO_3$ with Ln = La, Pr, Nd, Gd, and Dy (adapted from Ref. [19])

$Ln_{0.5}Ba_{0.5}CoO_3$ perovskite, the data were considered for fixed $\langle r_A \rangle$ = 1.196 Å as equal to that of $Dy_{0.5}Sr_{0.5}CoO_3$ perovskite and varied the σ^2 values for different perovskites (Fig. 3.13). The FM T_C increases with decreasing σ^2 and the T_C^0 value corresponding to the disorder-free case (σ^2 = 0.0) is around 217 K (inset Fig. 3.13). Therefore, for these type of perovskites it could be concluded that, increasing size disorder favors electronic phase separation, giving rise to magnetic clusters of different size ranges, while decrease in size disorder effect increases the FM/PM ratio.

Before concluding the discussion on complex magnetism, there is another important series of disordered perovskite cobaltites of the type $Ln_{1-x}Ca_xCoO_3$. These perovskites show no long-range ferromagnetism or insulator-metal transition, instead they exhibit electronic phase separation and/or glassy magnetic behavior at

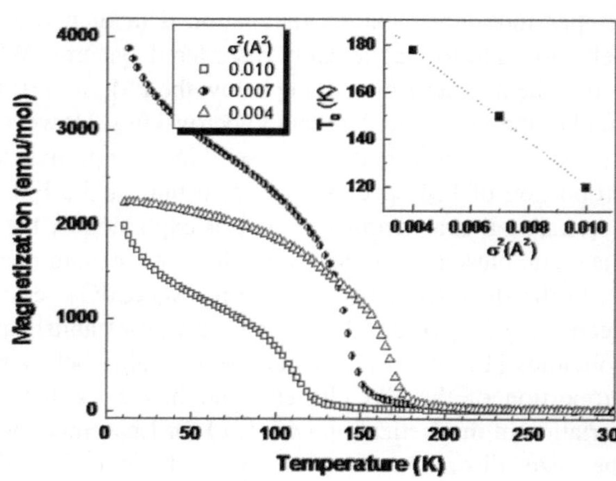

Fig. 3.13 Magnetic properties for a fixed $\langle r_A \rangle$ of 1.196 Å (adapted from Ref. [19])

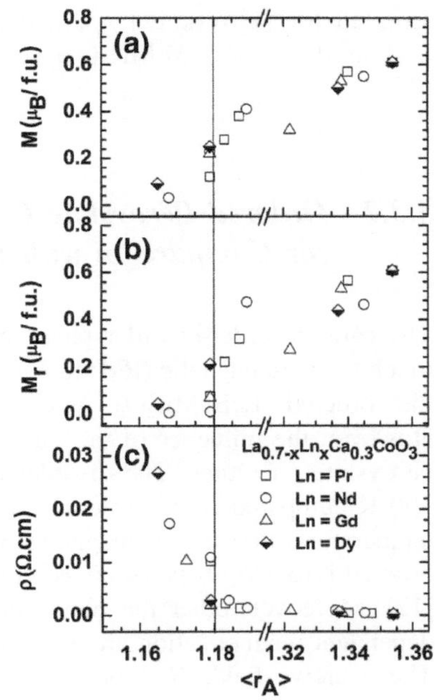

Fig. 3.14 Magnetization and Resistivity as function of $\langle r_A \rangle$ for disordered $La_{0.7-x}(Pr/Nd/Gd/Dy)_xCa_{0.5}CoO_3$ perovskite series (adapted from Ref. [11])

low temperatures [1, 11, 13]. Studies on $La_{1-x}Ca_xCoO_3$ perovskites have suggested that there are no major differences from the Sr-doped perovskites discussed in the previous section. Ferromagnetism is observed in both the perovskites, with the FM T_C being lower for Ca-substituted perovskites compared to Sr/Ba-substitution. Magnetic and electrical properties for different rare earth perovskites have been investigated to examine the effect of $\langle r_A \rangle$, and σ^2 on these perovskites as presented in Fig. 3.14. The perovskite $La_{0.7}Ca_{0.3}CoO_3$ ($\langle r_A \rangle$, = 1.354 Å) shows glassy ferromagnetism associated with metallicity at low temperature. Whereas, $Ln_{0.7}Ca_{0.3}CoO_3$ with a smaller $\langle r_A \rangle$, of 1.179 Å (Ln = *Pr*) and 1.168 Å (Ln = Nd) shows no long-range ferromagnetism or insulator–metal transition [11]. They exhibit electronic phase separation and/or spin glass-like behavior at low temperatures. The electronic phase separation and associated magnetic properties of $Pr_{0.7}Ca_{0.3}CoO_3$ and $Nd_{0.7}Ca_{0.3}CoO_{2.95}$ perovskite cobaltites arise because of the small average size of the A-site cations [11]. In these perovskites, the average radius (for orthorhombic structure) is less than 1.18 Å, which is the critical value only above which long-range ferromagnetism appeared. A detailed study on disordered rare earth perovskite cobaltites has shown the occurrence of electronic phase separation and glassy magnetic behavior for small $\langle r_A \rangle$, and a large σ^2 value [11].

In view of interrelation between phase separation and MR in the perovskite oxides, it is noticed that the value of MR for disordered perovskite cobaltites (even at T_C/T_{IM}) is much smaller in magnitude than for the perovskite manganites [7]. However, in disordered perovskite cobaltites the large and negative MR has been reported for perovskite $La_{0.5}Sr_{0.5}CoO_3$ in the insulating phase [7]. In this perovskite, the maximum MR is observed where the system shows *SG*-like behavior [7, 13].

3.3.2 Ordered Perovskite Cobaltites: A Special Case for Comparison with La–Ba–Co Disordered Phase

The cationic order-disorder phenomena in the perovskite cobaltites do not affect much the paramagnetic (PM) to ferromagnetic (FM) transition T_C [9], in contrast to the ordered $LaBaMn_2O_6$ and disordered $La_{0.5}Ba_{0.5}MnO_3$ manganites [20]. However, the influence of the cationic ordering on the T_C seems to be reverse for perovskite cobaltites. The disordered perovskite cobaltites exhibit a higher T_C of 190 K compared to the ordered phase ($T_C \sim 175$ K) [9]. Figure 3.15 shows the temperature dependent magnetization for the disordered $La_{0.5}Ba_{0.5}CoO_3$ and ordered $LaBaCo_2O_6$ perovskites with the T_C values of 177 and 174 K, respectively [12]. Moreover, other magnetic behaviors such as the field and the temperature dependent magnetization are expected to be similar as for PM/FM phases [9, 12]. The coercive field, H_C, for disordered $La_{0.5}Ba_{0.5}CoO_3$ and ordered $LaBaCo_2O_6$ perovskites are 0.08 and 0.05 (Tesla) T, respectively (at 10 K); the low value of H_C signifies the nature of soft FM material [12].

There is another perovskite of Nd–Ba–Co reported in the literature as ordered-disordered perovskite phases, with a FM T_C of 200 K in the ordered

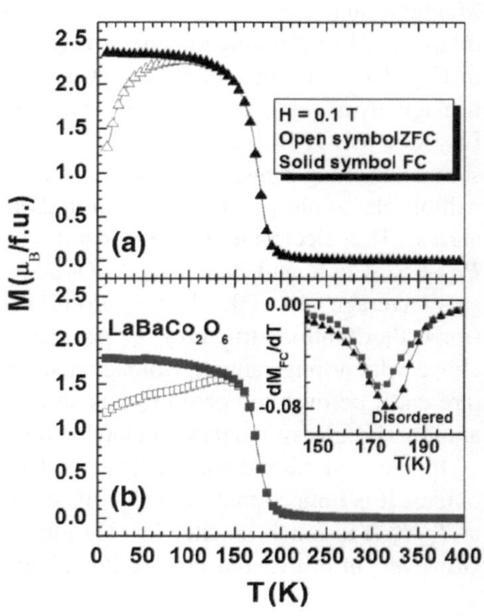

Fig. 3.15 Temperature dependence ZFC (*open symbol*) and FC (*solid symbol*) Magnetization ($H = 0.1$ T) for **a** disordered $La_{0.5}Ba_{0.5}CoO_3$ and **b** ordered $LaBaCo_2O_6$. The *inset* figure shows dM_{FC}/dT versus temperature plot (adapted from Ref. [12])

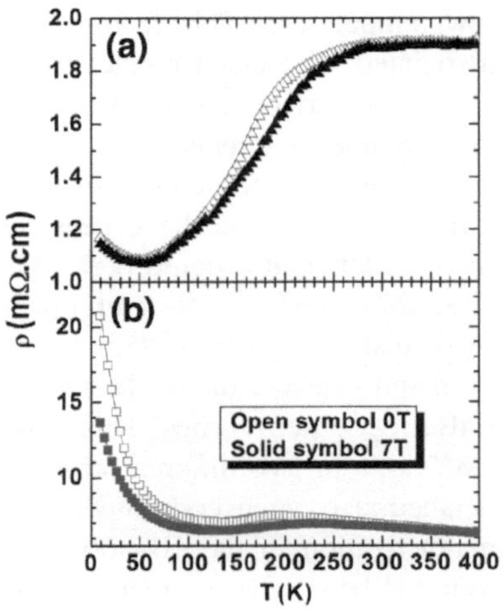

Fig. 3.16 Temperature dependence electrical resistivity for **a** disordered $La_{0.5}Ba_{0.5}CoO_3$ and **b** ordered $LaBaCo_2O_6$ cobaltites (adapted from Ref. [12])

$NdBaCo_2O_6$ phase [14] whereas for disordered $Nd_{0.5}Ba_{0.5}CoO_3$ phase [13] the value is 130 K. Likewise, the ordered perovskite *Pr*$BaCo_2O_6$ with a FM T_C of 210 K by Seikh et al. [14], which is higher than those of ordered perovskites $LaBaCo_2O_6$ ($T_C \sim 179$ K) and $NdBaCo_2O_6$ ($T_C \sim 200$ K), suggesting that the size of the lanthanide influences the double exchange mechanism as well as the FM T_C of these perovskites. Since, the perovskite cobaltites Nd–Ba–Co are reported by two different groups along with separate characterizations so we will not elaborate our discussion on those ordered-disordered perovskites for the present book.

Figure 3.16 shows the electrical resistivity with the variation of temperature for both the ordered-disordered La–Ba–Co perovskites in an applied magnetic field of ±7 T. The resistivity, $\rho(T)$, behavior of the disordered $La_{0.5}Ba_{0.5}CoO_3$, and ordered $LaBaCo_2O_6$ is distinctly different in 10–400 K temperature range, which depicts semi-metallic behavior at high temperature ($T > 300$ K) for the disordered $La_{0.5}Ba_{0.5}CoO_3$ (Fig. 3.16a), whereas the ordered $LaBaCo_2O_6$ is semiconducting down to 190 K (Fig. 3.16b). This is due to the 180° ⟨Co–O–Co⟩ bond angle of disordered $La_{0.5}Ba_{0.5}CoO_3$ in this temperature range, which is in agreement with the cubic or pseudo cubic structure, favoring a perfect overlapping of the Co 3d orbitals and oxygen 2*p* orbitals. But, for the ordered $LaBaCo_2O_6$ perovskite the ⟨Co–O–Co⟩ bond angles of 174° in the equatorial planes of the $[CoO_2]_\alpha$ layers are observed at room temperature [9]. Hence for the ordered perovskite the conduction of charge carriers is more favorable for linear bond angle, as a result metallic type of conductivity is noticed for the disordered perovskite $La_{0.5}Ba_{0.5}CoO_3$. With decreasing temperature a transition to a nearly metallic state is observed for both the perovskites. This is characterized by a change in slope of $\rho(T)$ at T_C for the disordered $La_{0.5}Ba_{0.5}CoO_3$ (Fig. 3.16a), or by a flat maximum at T_C for the ordered $LaBaCo_2O_6$ (Fig. 3.16b). These results show that irrespective of the structural nature, the different forms of perovskite cobaltites exhibit a FM metallic behavior below T_C. Moreover, both perovskites depict an upturn in the resistivity behavior at low temperature. This feature is interpreted as a weak localization contribution associated with electron–electron interaction. In the present case, the magnetoresistance behavior that is discussed later, suggest that the upturn is rather due to grain boundary effects.

It is now important to discuss about the oxygen deficient ordered cobaltites $LaBaCo_2O_{5.5}$, which have been derived from ordered cobaltite $LaBaCo_2O_6$. This oxygen deficient ordered perovskite is of great interest due to rich physical properties and interesting structural phenomena. In ordered $LaBaCo_2O_{5.5}$ perovskite the oxygen stoichiometry is "5.5," hence the cobalt valency is Co^{3+} unlike the presence of mixed valences of cobalt in the other two disordered $La_{0.5}Ba_{0.5}CoO_3$ and ordered $LaBaCo_2O_6$ perovskites. This is particularly interesting because of the ordering of Co^{3+} ions in two different crystallographic sites corresponding to pyramidal and octahedral oxygen coordination as discussed earlier. The temperature dependent magnetization in the 10–400 K temperature range is presented in Fig. 3.17 are found to be similar to other lanthanide perovskitess [3, 14]. The perovskite exhibits several magnetic transitions from PM to FM-like to antiferromagnetic with the decreasing temperature from 400 to 10 K. The rapid increase of magnetization around 326 K indicates a PM/FM transition T_C and a sharp drop in the magnetization around 295 K as a FM/AFM transition T_N. The nature of magnetic interactions between Co^{3+} ions, both in pyramidal and octahedral coordination of cobalt,

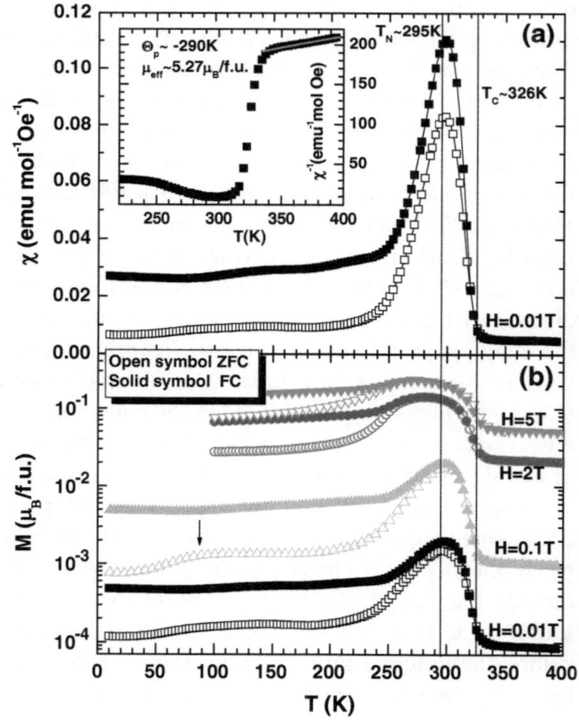

Fig. 3.17 Temperature dependent ZFC (*open symbol*) and FC (*solid symbol*) Magnetization behavior of ordered $LaBaCo_2O_{5.5}$; **a** magnetic susceptibility, χ, under $H = 0.01$ T (*inset* figure shows the inverse magnetic susceptibility, χ^{-1}, versus temperature plot and solid line is Cuire–Weiss fitting) and **b** Magnetic moment in different magnetic fields (adapted from Ref. [16])

are found to be AFM at low temperature. Additionally, the divergence between ZFC and FC magnetization at low temperature AFM region remains well discernible even higher fields. This is due to the fact that, a high field intrinsically affects the FM-AFM competition, which in turn suppresses the magnetization drop below the T_N.

The nonzero magnetization in the AFM state down to 10 K signifies the presence of some kind of FM-like interactions, where some weak magnetic transition is also depicted. Therefore, in the presence of external magnetic field, the FM state becomes more stable (indeed T_C increases and the FM region expands in the temperature scale), however, the T_N shifts to low temperature.

The inverse susceptibility versus temperature plot for ordered $LaBaCo_2O_{5.5}$ perovskite is shown in the inset of Fig. 3.17, which follows the Curie–Weiss behavior with PM Weiss temperature (θ_p) of ~−290 K and an effective PM moment (μ_{eff}) of 5.27 μ_B/f.u. The large negative θ_p value for ordered $LaBaCo_2O_{5.5}$ perovskite indicates the existence of AFM type interactions in the high temperature region. The μ_{eff} value of 5.27 μ_B/f.u. indicates a situation where the Co^{3+} ions are in the *IS* state [16]. The sharp drop in inverse susceptibility near T_C, is similar to that observed for other perovskite cobaltites $YBaCo_2O_{5.5}$ and $GdBaCo_2O_{5.5}$ [21, 24].

The FM-like features below T_C ~ 326 K, is confirmed by the isotherm magnetization, *M*(*H*), as shown in Fig. 3.18 at different temperatures. At room temperature, the *M*(*H*) curve shows a prominent hysteresis loop with a remanent magnetization, M_r, and a coercive field, H_C, values of 0.02 μ_B/f.u and 0.1 T, respectively, indicates a FM state below T_C. However, the highest value of magnetic moment of 0.23μ_B/f.u. at 275 K is relatively smaller than the theoretical spin-only value of Co^{3+} in IS state (4μ_B/f.u.). The FM-like behavior of ordered $LaBaCo_2O_{5.5}$ perovskite is due to the canting of the magnetic spin alignment in the AFM phase, often observed in other perovskites [14]. Although there are some

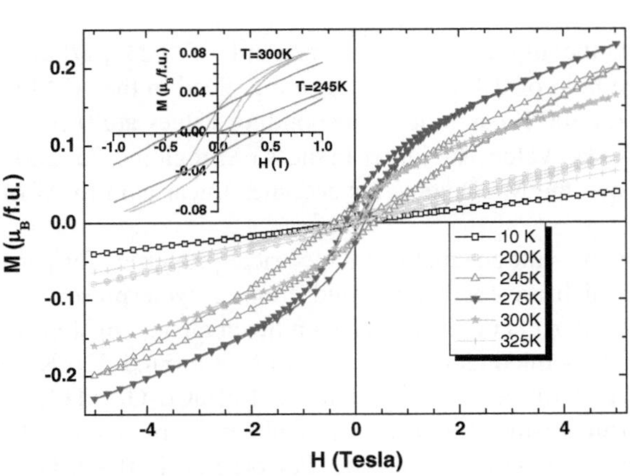

Fig. 3.18 Magnetic field dependent isotherm magnetization, *M*(*H*), of ordered $LaBaCo_2O_{5.5}$ at six different temperatures. The *inset* figure shows the expanded version for lower magnetic fields at 245 and 300 K (adapted from Ref. [27])

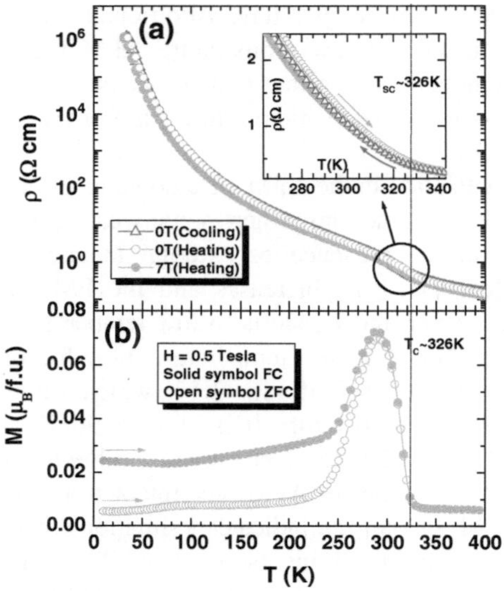

Fig. 3.19 Temperature dependent physical properties for ordered $LaBaCo_2O_{5.5}$. **a** Electrical resistivity, $\rho(T)$, in the presence (*solid symbol*) and absence (*open symbol*) of magnetic field (7 T) during heating and cooling cycles (*inset* shows the expanded version near the transition temperature, T_{IM}), and **b** ZFC (*open symbol*), FC (*solid symbol*) Magnetization in an applied field of 0.5 Tesla (adapted from Ref. [27])

controversies in the literature to explain these FM-like features, yet this behavior is prominent for this ordered $LaBaCo_2O_{5.5}$ perovskite. It is also important to mention that in the low temperature AFM region, some fractions of FM-like phase is present, which are evidenced from the *M*(*H*) behavior, with a finite value of the coercive field.

Interestingly, the highest magnetic moment of 0.23 μ_B/f.u. (at 275 K) and highest coercive field of 0.4 T (at 245 K) are noticed in the AFM region instead in the FM region (at 300 K) and the corresponding values are 0.16 μ_B/f.u. and 0.1 T, respectively. The H_C values are larger in the AFM region (275–200 K) compared to the FM region and the M(H) behavior becomes linear akin to AFM state finally at 10 K.

Figure 3.19 shows the magnetic and electrical properties for perovskite cobaltite $LaBaCo_2O_{5.5}$, which are for heating and cooling cycle presence of the magnetic field. The zero field $\rho(T)$ curve shows a significant change in slope corresponding to the semiconductor-semiconductor transition (T_{SC}) around 326 K (Fig. 3.19a) similar to other series of perovskite cobaltites $LnBaCo_2O_{5.5}$ [14, 24, 25]. This is referred to as insulator-metal transition T_{IM} albeit the true nature of this transition is semiconducting to semiconducting type. For ordered $LaBaCo_2O_{5.5}$ perovskite, the slope of the resistivity ($d\rho/dT$) is negative above $T > T_{SC}$, in contrast to a metallic behavior. It is also noticed that, the electronic and magnetic transition temperatures for $LaBaCo_2O_{5.5}$ perovskite are almost same, in contrast to other ordered

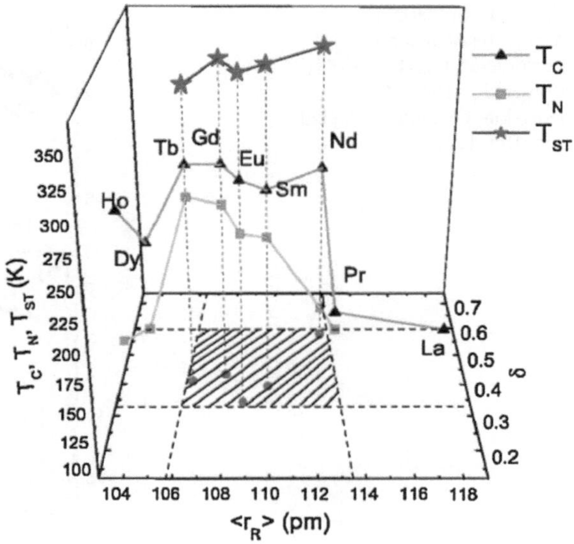

Fig. 3.20 Variation of T_C and T_N for ordered $LnBaCo_2O_{5+\delta}$ perovskites (Ln = rare earths) (adapted from Roy et al. [14])

$LnBaCo_2O_{5.5}$ perovskites, where large shifts are evidenced as shown in Fig. 3.20 [14, 24–26].

The semiconducting or insulating like transport mechanism in perovskite cobaltites have been analyzed by three possible models [11, 27] namely thermal activation (TA): log ρ α $1/T$, Efros–Shklovskii type hopping (ESH): log ρ α $T^{-1/2}$ and Mott's variable range hopping (*VRH*): log ρ α $T^{-1/4}$. The zero field resistivity data for ordered $LaBaCo_2O_{5.5}$ perovskite, shows that the *VRH* behavior, and is consistent with the preceding studies on perovskite cobaltites [11, 27]. Taskin et al. [25] have described the formation of localized states in terms of oxygen defects in ordered $GdBaCo_2O_{5.5}$ perovskite, which certainly generate electrons or holes in the CoO_2 planes.

In the final part of discussion, we have also considered it important to compare the magnetoresistance (MR) behavior of the stoichiometric ordered-disordered perovskites. Figure 3.21 shows the magnetic field dependent resistivity behavior at low temperature and/or below T_C. The MR value generally calculated as MR (%) = [{$\rho(7) - \rho(0)$}/$\rho(0)$)] × 100, where $\rho(0)$ is resistivity at zero field and $\rho(7)$ under the magnetic field of ±7 T. For the disordered $La_{0.5}Ba_{0.5}CoO_3$ perovskite (Fig. 3.21a), the maximum MR of 7 % is observed around T_C, whereas for ordered $LaBaCo_2O_6$ perovskite the corresponding value is 6 % (Fig. 3.21b). However, the ordered $LaBaCo_2O_6$ perovskite depicts an MR value up to 14.5 % at 10 K in an applied field of ±7 T which is much larger than an MR value of 4 % for the disordered $La_{0.5}Ba_{0.5}CoO_3$ perovskite (10 K) [12]. The large value of MR for the ordered $LaBaCo_2O_6$ perovskite suggests that, the grain boundary effect plays an important role in the anisotropic MR behavior at low temperature (T < 50 K). This

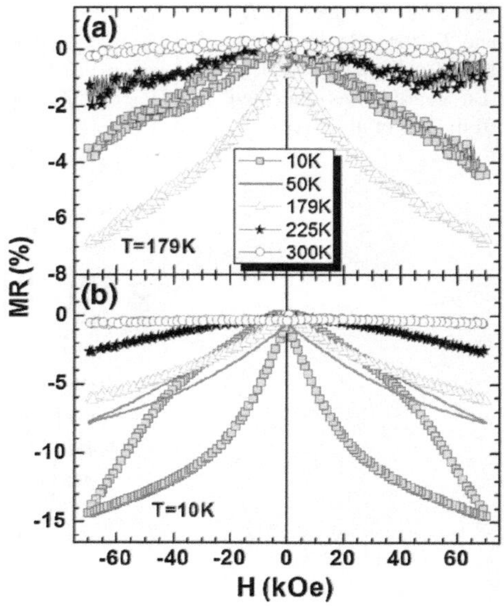

Fig. 3.21 Isothermal magnetoresistance for **a** disordered $La_{0.5}Ba_{0.5}CoO_3$ **b** ordered $LaBaCo_2O_6$ perovskite cobaltites (adapted from Ref. [12])

is in agreement with the rapid upturn of resistivity at low temperature, which is almost 10 times higher (at 10 K) than disordered $La_{0.5}Ba_{0.5}CoO_3$ perovskite. Hence, the higher MR value for the ordered $LaBaCo_2O_6$ perovskite is due to the tunneling magnetoresistance (TMR) effect which increase with the intergrain insulating barriers [26], rather than an intrinsic effect. This effect is dominant at 10 K over the TMR effect for the disordered $La_{0.5}Ba_{0.5}CoO_3$ perovskite, which is due to the spin-polarized tunneling of carriers across the insulating boundaries. They occur at the interfaces between polycrystalline grains as a result the TMR effect in these perovskites noticed.

Similarly, for ordered $LaBaCo_2O_{5.5}$ perovskite the MR is noticed below the T_C, Fig. 3.22 shows the MR effect at five different temperatures. As discussed previously, the temperature dependent $\rho(T)$ curve shows no major change even though the magnetic behavior signifies the presence of FM and AFM ordering [27]. The charge transport for this perovskite is expected to be very sensitive due to the coexistence of FM and AFM state. As a result, the external magnetic field induces an MR in the perovskite by affecting the subtle balance between FM-AFM phases.

In contrast to the disordered $La_{0.5}Ba_{0.5}CoO_3$ and ordered $LaBaCo_2O_6$ perovskites, the highest MR value of 5 % is observed at 245 K and at 300 K value is only 1.6 %. The negative MR value at low temperature, akin to the TMR observed usually in polycrystalline samples, in the case of ordered $LaBaCo_2O_{5.5}$, is associated to the spin-dependent scattering at grain boundaries. Nevertheless, for the $LaBaCo_2O_{5.5}$ perovskite, the highest MR value is noticed near the FM-AFM phase boundary, and hence the grain boundary effect could be neglected and considered to be as an intrinsic property of the perovskite. Interestingly, the isothermal MR behavior at 245 K exhibits an irreversible effect analogous to those of isothermal

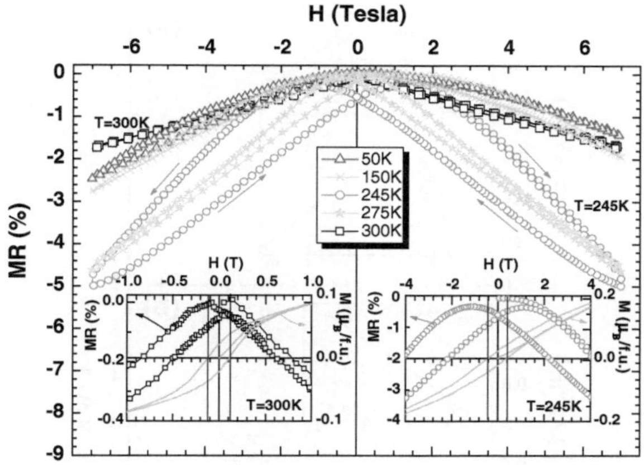

Fig. 3.22 Magnetic field dependent isotherm magnetoresistance, MR, effect for ordered $LaBaCo_2O_{5.5}$ at five different temperatures (H = ±7 T). *Inset* figures show the isotherm magnetization, *M*(*H*), and MR plot at 245 and 300 K for comparison (adapted from Ref. [27])

magnetization, *M*(*H*), behavior (inset Fig. 3.22), also observed for 300 K. The peak value in isothermal MR occurs at the coercive field value, which corresponds to the state of maximum disorder in the orientation of the neighboring spins. Thus, the field dependent MR data that is indirectly related to the alignment between magnetic spins, reaches a maximum value. This effect is prominent for 300 K, compared to 245 K as shown in the inset of Fig. 3.22. This is due to FM-like state near 300 K whereas the 245 K corresponds to magnetic phase boundary. The isothermal MR effects resemble the "butterfly-like" feature, although the effect is rather weak at low temperature with strong irreversible nature near the FM-AFM phase boundary. Correspondingly, the magnetic field dependent isotherm MR behavior at 10 K for the ordered-disordered perovskites exhibits an anisotropic effect similar to the magnetization behavior. Nevertheless, the isothermal MR behavior below room temperature for all three perovskites exhibit hysteresis effects, which resemble the "butterfly-like" feature, although the effect is relatively weak for the disordered $La_{0.5}Ba_{0.5}CoO_3$ perovskite. It is noticed that the MR effect is nearly isotropic near or above the T_C. Therefore, the butterfly-like curve appears only at low temperatures, which is prominent for ordered $LaBaCo_2O_{5.5}$ and $LaBaCo_2O_6$ perovskites. The origin of isothermal MR near the T_C for disordered $La_{0.5}Ba_{0.5}CoO_3$ and ordered $LaBaCo_2O_6$ perovskites are explained by the mechanism of suppression of spin fluctuations at low temperature. However, the highest MR value at 10 K for the ordered $LaBaCo_2O_6$, is explained by the TMR effect due to the presence of more insulating grain boundaries. Hence, the appearance of irreversible MR behavior nearly at similar temperatures for magnetic field dependence of isothermal M(H) suggest the strongly correlated nature of magnetic field-induced magnetic and electronic transitions [12, 27].

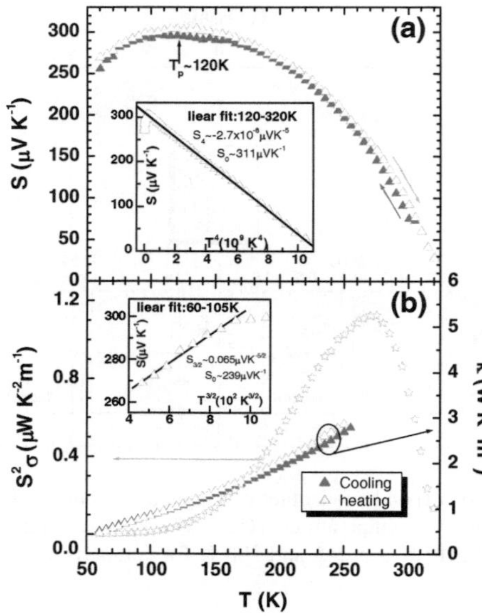

Fig. 3.23 Temperature dependent transport measurements for ordered perovskite $LaBaCo_2O_{5.5}$. **a** Thermoelectric power, $S(T)$, during cooling (*solid triangle*) and heating (*open triangle*) cycles and *inset* shows the S–T^4 plot in the 120–320 K range, **b** thermal conductivity, $\kappa(T)$, and power factor, $S^2\sigma(T)$, in the temperature range of 60–320 K (*inset* shows the S–$T^{3/2}$ plot in the 60–105 K range) (adapted from Ref. [27])

For ordered $LaBaCo_2O_{5.5}$ perovskite we have also discussed the thermopower, $S(T)$, and thermal conductivity, $\kappa(T)$, measurements in order to explain the conduction mechanism below T_{SC}. These are influenced by the magnetic and electrical nature of charge carriers (hole/electron), which are absent in the magnetotransport behavior. Moreover, the $S(T)$ is less affected by the grain boundaries, which often complicates the $\rho(T)$ behavior interpretation for polycrystalline samples. The $LaBaCo_2O_{5.5}$ perovskite shows relatively large positive value of the thermoelectric power (91 μV/K) at 300 K as shown in Fig. 3.23.

The $S(T)$ behavior is positive below T_{SC} and the value increases gradually with decreasing temperature and reaches a maximum value of 303 μV/K around 120 K. The $S(T)$ decreases rapidly to lower value with further cooling and the behavior is similar to other reported perovskites i.e. $NdBaCo_2O_{5.5}$, $GdBaCo_2O_{5.5}$ and $Ho$$BaCo_2O_{5.5}$ respectively [25, 28]. The $S(T)$ behavior signifies that the $LnBaCo_2O_{5.5}$ perovskites show a semiconducting type behavior of the thermopower down to low temperature similar to resistivity behavior. In contrast, with decreasing temperature the $S(T)$ value was expected to increase due to trapping or localization of charge carriers. This type of $S(T)$ behavior for semiconducting thermoelectric materials is not common. We have discussed the general approach to analyze the semiconducting behavior and plotted the $S(T)$ in the $T^{-1/n}$ scale similar to $\rho(T)$. For semiconductors the $S(T)$ is expected to be linear in T^{-1} (TA model) or follow the earlier described hopping models similar to $\rho(T)$ [11, 27]. The $S(T)$ behavior could not be described by the hopping models. The thermopower data analyzed by the expression $S(T) = S_0 + S_{3/2}T^{3/2} + S_4T^4$ proposed by *P*. Mandal [29], which could be explained on the basis of electron-magnon scattering (spin wave theory). The $S(T)$ data in 60 K $\leq$ T $\leq$ 105 K range follows $T^{3/2}$ behavior and in the 120–320 K range, it fits linearly with the T^4 behavior (insets of Fig. 3.23). At low temperature the second term ($S_{3/2}$) dominate over S_4 ($S_{3/2} \gg S_4$), hence the $S(T)$ will depict downward trend at low temperature. The broad peak of $S(T)$ data linearly fit to the $T^{3/2}$ term (inset of Fig. 3.23b), as expected from the spin wave theory. Importantly, the downward trend of $S(T)$ in all reported $LnBaCo_2O_{5.5}$ perovskites could also be explained by this theory [25, 27, 28]. Therefore, the broad peak at low temperature and downward trends for $LnBaCo_2O_{5.5}$ perovskites are due to the electron magnon scattering similar to the perovskite manganite [29].

References

1. C.N.R. Rao, B. Raveau (Eds.), *Colossal Magnetoresistance, Charge-Ordering and Related Properties of Manganese Oxides* (World Scientific, Singapore, 1998); Dagotto E., *Nanoscale Phase Separation and Colossal Magnetoresistance: The Physics of Manganites and Related Compounds* (Springer, Berlin, New York, 2003); C.N.R. Rao, A.K. Kundu, M.M. Seikh, L. Sudheendra, Dalton Trans. **19**, 3003 (2004); B. Raveau, M.M. Seikh, *Cobalt Oxides: From Crystal Chemistry to Physics* (Wiley-VCH, 2012)
2. G.H. Jonker, J.H. van Santen, Physica **19**, 120 (1953); J.B. Goodenough, J. Phys. Chem. Solids **6**, 287 (1958); G.H. Jonker, J. Appl. Phys. **37**, 1424 (1966)
3. C. Martin, A. Maignan, P. Pelloquin, N. Nguyen, B. Raveau, Appl. Phys. Lett. **71**, 1421 (1997); A. Maignan, C. Martin, D. Pelloquin, N. Nguyen, B. Raveau, J. Solid State Chem. **142**, 247 (1999)
4. Y. Teraoka, T. Nobunaga, K. Okamoto, N. Miura, N. Yamazoe, Solid State Ionics **48**, 207 (1991); H. Kruidhof, H.J.M. Bouwmeester, R.H.E. van Doorn, A. Burggraaf, Solid State Ionics **63**, 816 (1993); R.H.E. van Doorn, A. Burggraaf, Solid State Ionics **128**, 65 (2000); G. Kim, S. Wang, A.J. Jacobson, L. Reimus, P. Brodersen, C.A. Mims, J. Mater. Chem. **17**, 2500 (2007)
5. K. Takada, H. Sakurai, E. Takayama-Muromachi, F. Izumi, R. Dilanian, T. Sasaki, Nature **422**, 53 (2003)
6. M. Itoh, I. Natori, S. Kubota, K. Matoya, J. Phys. Soc. Jpn. **63,** 1486 (1994); P.S. Anil Kumar, P.A. Joy, S.K. Date, J. Phys.: Condens. Matter **10**, L487 (1998); I.O. Troyanchuk, N.V. Kasper, D.D. Khalyavin, H. Szymczak, R. Szymczak, M. Baran, Phys. Rev. Lett. **80**, 3380 (1998); D.N.H. Nam, K. Jonason, P. Nordblad, N.V. Khiem, N.X. Phuc, Phys. Rev. B, **59**, 4189 (1999); P.L. Kuhns, M.J.R. Hoch, W.G. Moulton, A.P. Reyes, J. Wu and C. Leighton, Phys. Rev. Lett. **91**, 127202 (2003); J.C. Burley, J.F. Mitchell, S. Short, Phys. Rev. B **69**, 054401 (2004)
7. R. Mahendiran, A.K. Raychaudhuri, A. Chainani and D.D. Sarma, J. Phys.: Condens. Mater. **7**, L561 (1995); R. Mahendiran, A.K. Raychaudhuri, Phys. Rev. B **54**, 16044 (1996)
8. F. Fauth, E. Suard, V. Caignaert, Phys. Rev. B **65**, 60401 (2001)
9. T. Nakajima, M. Ichihara, Y. Ueda, J. Phys. Soc. Jpn. **74**, 1572 (2005)

10. J. Wu, J.W. Lynn, C.J. Glinka, J. Burley, H. Zheng, J.F. Mitchell, C. Leighton, Phys. Rev. Lett. **94**, 037201 (2005); D. Fuchs et al., Phys. Rev. B, **71**, 92406 (2005); M.W. Haverkort et al., Phys. Rev. Lett., **97**, 176405 (2006); J. Yu et al., Phys. Rev. B, **80**, 052402, (2009); C. He et al., Phys. Rev. B, **80**, 214411, (2009); D. Phelan et al., Phys. Rev. B, **89**, 184427, (2014)
11. M. Respaud et al., Phys. Rev. B **64**, 214401 (2001); F. Fauth et al., Phys. Rev. B **66**, 184421, (2002); D.D. Khalyavin et al., Phys. Rev. B **67**, 214421, (2003); Z. X. Zhou et al., Phys. Rev. B, **70**, 24425, (2004); A. K. Kundu, K. Ramesha, R. Seshadri, C.N.R. Rao, J. Phys. Condens. Matter **16**, 7955 (2004); A.K. Kundu, E.V. Sampathkumaran, K.V. Gopalakrishnan and C.N. R. Rao, J. Mag. Mag. Mater. **281**, 261 (2004); A.K. Kundu, P. Nordblad, C.N. R. Rao, Phys. Rev. B **72**, 144423 (2005); V.P. Plakhty et al., Phys. Rev. B, **71**, 214407, (2005); B. Raveau et al., J. Phys. Condens. Matter **18**, 10237, (2006); B. Raveau et al., Solid State Commun. **139**, 301, (2006); G. Aurelio et al., Physica B, **384**, 106, (2006); G. Aurelio et al., Phys. Rev. B **76**, 214417 (2007); M García-Fernández et al., Phys. Rev. B **78**, 054424, (2008); M. M. Seikh et al., Solid State Commun., **149**, 697 (2009); T. Sarkar et al., Phys. Rev. B, **83**, 214428 (2011); J. Wieckowski et al., Phys. Rev. B **88**, 054404 (2012)
12. E.-L. Rautama, P. Boullay, A.K. Kundu, V. Pralong, V. Caignaert, M. Karppinen, B. Raveau, Chem. Mat. **20**, 2742 (2008)
13. C.N.R. Rao, V.G. Bhide, N.F. Mott, Phil. Mag. **32**, 1277 (1975); V.G. Bhide, D.S. Rajoria, C. N.R. Rao, G.R. Rao, V.G. Jadhao, Phys. Rev. B **12**, 2832 (1975); C.N.R. Rao, O. Prakash, D. Bahadur, P. Ganguly and S. Nagabhushana, J. Solid State Chem. **22**, 353 (1977); Y. Moritomo, M. Takeo, X. J. Liu, T. Akimoto, A. Nakamura, Phys. Rev. B **58**, R13334 (1998); R. Ganguly, A. Maignan, C. Martin, M. Hervieu and B. Raveau, J. Phys.: Condens. Matter **14**, 8595 (2002)
14. Y. Moritomo, T. Akimoto, M. Takeo, A. Machida, E. Nishibori, M. Takata, M. Sakata, K. Ohoyama, A. Nakamura, Phys. Rev. B **61**, R13325 (2000); S. Roy, M. Khan, Y. Q. Guo, J. Craig, N. Ali, Phys. Rev. B **65**, 064437 (2002); S. Roy, I.S. Dubenko, M. Khan, E.M. Condon, J. Craig, N. Ali, Phys. Rev. B **71**, 024419 (2005); V. Pralong, V. Caignaert, S. Hebert, A. Maignan, B. Raveau, Solid State Ionics **177**, 1879 (2006); M.M. Seikh, C. Simon, V. Caignaert, V. Pralong, M.B. Lepetit, S. Boudin and B. Raveau, Chem. Mat. **20**, 231 (2008); M.M. Seikh, V. Pralong, O.I. Lebedev, V. Caignaert, B. Raveau, J. Appl. Phys. **114**, 013902 (2013)
15. T. Vogt, P.M. Woodward, P. Karen, B.A. Hunter, P. Henning, A.R. Moodenbaugh, Phys. Rev. Lett. **84**, 2969 (2000); E. Suard, F. Fauth, V. Caignaert, I. Mirebeau, G. Baldinozzi, Phys. Rev. B **61**, R11871 (2000)
16. E.L. Rautama, V. Caignaert, P. Boullay, A.K. Kundu, V. Pralong, M. Karppinen, C. Ritter, B. Raveau, Chem. Mat. **21**, 102 (2009)
17. L.M. Rodriguez-Martinez, J.P. Attfield, Phys. Rev. B **54**, R15622 (1996); J. P. Attfield. Chem. Mater. **10**, 3239 (1998)
18. A.K. Kundu, C.N.R. Rao, J. Phys.: Condens. Matter **16**, 415 (2004); A.K. Kundu, E.V. Sampathkumaran, C.N.R. Rao, J. Phys. Chem. Solids **65**, 95 (2004)
19. A.K. Kundu, R. Sarkar, B. Pahari, A. Ghoshray, C.N.R. Rao, J. Solid State Chem. **180**, 1318 (2007). and references therein
20. F. Millange, V. Caignaert, B. Domengès, B. Raveau, E. Suard, Chem. Mater. **10**, 1974 (1998)
21. D. Akahoshi, Y. Ueda, J. Solid State Chem. **156**, 355 (2001)
22. M. Abbate, J.C. Fuggle, A. Fujimori, L.H. Tjeng, C.T. Chen, R. Potze, G.A. Sawatzky, H. Eisaki, S. Uchida, Phys. Rev. B **47**, 16124 (1993); M.A. Korotin, S.Y. Ezhov, I.V. Solovyev, V.I. Anisimov, D.I. Khomskii, G.A. Sawatzky, Phys. Rev. B **54**, 5309 (1996); H. Wu, Phys. Rev. B **62**, R11953 (2000)
23. C. Zener, Phys. Rev. **82**, 403 (1951); J.B. Goodenough, A. Wold, R.J. Arnott, M. Menyuk, Phys. Rev. **124**, 373 (1961)
24. A.A. Taskin, A.N. Lavrov, Y. Ando, Phys. Rev. Lett **90**, 227201 (2003)

25. A.A. Taskin, A.N. Lavrov, Y. Ando, Phys. Rev. B **71**, 134414 (2005); A.A. Taskin, A.N. Lavrov, Y. Ando, Phys. Rev. B **73**, R121101 (2006)
26. D. Niebieskikwiat, F. Prado, A. Caneiro, R.D. Sanchez, Phys. Rev. B **70**, 132412 (2004)
27. A.K. Kundu, B. Raveau, V. Caignaert, E.-L. Rautama, V. Pralong, J. Phys.: Condens. Matter **21**, 056007 (2009)
28. A. Maignan, V. Caignaert, B. Raveau, D. Khomskii, G. Sawatzky, Phys. Rev. Lett. **93**, 26401 (2004)
29. P. Mandal, Phys. Rev. B **61**, 14675 (2000)

Chapter 4
Bismuth-Centered Perovskite Multiferroics

4.1 Introduction

In the last few decades there has been an increasing interest in the understanding of the basic physics/chemistry of multiferroics and/or spintronics materials [1–13]. The interplay between the structural, magnetic, and electronic properties gives rise to fascinating complex phenomena and therefore the basic physics of the materials is rich. More specifically, in the transition metal oxides strong interplay between lattice, charge, spin and/or orbital degrees of freedom provide a fantastic playground to tune their physical properties. In this respect, the ferromagnetic semiconductors/insulators are novel singular materials that could exhibit simultaneously electric and magnetic ordering [2, 7–15]. They also have practical applications in spintronics and magnetodielectric-based devices such as nonvolatile memories, magnetic read heads, tunnel junction spin filtering, etc. [1–10, 16]. Also, the recent investigations on multiferroic and/or spin filtering in this type of thin films have enhanced the possibility of device applications [16, 17]. Despite the numerous investigations on spintronics materials in the past few years a very few perovskites have been known to realize as ferromagnetic insulators (FMI) [6–15, 18–26].

Multifunctional materials (multiferroics, magnetodielectric, spintronics, etc.) have attracted increasing attention due to their possible applications toward storage materials and intriguing fundamental physics [1–10, 16]. Among the naturally existing oxides, the presence of both ferromagnetism and ferroelectricity is a rare phenomenon, due to the incompatibility between magnetism and ferroelectricity [2]. This incongruity could be at the origin of a limited number of multiferroics, though the researchers are looking for such materials from more than six decades. This phenomenon often occurs in perovskite oxide having the general formula ABO_3. In the process of exploration of a multiferroic perovskite the following facts are now well established: (i) ferromagnetic (FM) and ferroelectric (FE) behaviors are mutually exclusive due to the d^0 electronic structure of the *B*-site element [2], (ii) the occupation of different *B*-site cations with varying ionic radius provides an

A.K. Kundu, *Magnetic Perovskites*, Engineering Materials,
DOI 10.1007/978-81-322-2761-8_4

opportunity to realize a polar ground state [12], and (iii) the lattice distortion induced by cations with lone pair electrons such as Pb^{2+} or Bi^{3+} play a primordial role on the FE properties as reported for $PbTiO_3$ in comparison with $BaTiO_3$ [8].

The most well-known examples of existing perovskite multiferroics are $BiFeO_3$ and $BiMnO_3$ [3–5]. A larger number of investigations carried out on both these oxides revealed that $BiFeO_3$ is a so-called canted antiferromagnet, which gives rise to weak ferromagnetism and the $BiMnO_3$ is metastable, requiring high pressure conditions for synthesis of bulk phases [3, 5, 7, 8]. In the recent years a number of compounds have been reported to exhibit simultaneous electric and magnetic ordering in a similar way to that of the magnetoelectric or multiferroics behavior. A few examples are:

- $CdCr_2Se_4$ (Ref. [1]), $CaMn_7O_{12}$ (Johnson et al. [10]), $MnWO_4$ (Heyer et al. [10])
- $La_2Mn(Co/Ni)O_6$ (Refs. [9, 27]), $YBaCuFeO_5$ (Kundys et al. [10])
- Bi–Mn–(Ni/Fe/Cr)–O (Refs. [13, 28])
- $LnMnO_3$ (Goto et al., Kimura et al. and Aoyama et al. [10], Efremov et al. [4]), $(Sm/Ba)MnO_3$ (Sakai et al. and Pratt et al. [10]), $LnMn_2O_5$ (Lee et al. [10]),
- $LnFeO_3$ (Tokunga et al. and Saha et al. [10]), $GaFeO_3$ $LnFe_2O_4$ (Ikeda et al. and Rao et al. [10]), CuO (Kimura et al. [10]), Ln_2BaNiO_5 (Basu et al. [10])
- $LnCrO_3$ (Serrao et al., and Sahu et al. [10]),Cr_2O_3(Rado et al. [10])
- Diluted magnetic semiconductors [29].

The studies on $BiMn_{0.5}Ni_{0.5}O_3$ throw light on synthesis of materials with one or more order parameters for realizing multiferroic properties or magnetoelectric effects [13]. Most of the Bi^{3+} based ordered perovskites studied for multiferroic and magnetodielectric properties are recognized to exhibit high sensitivity toward the *B*-site cationic ordering. Unfortunately, they require high pressure conditions for synthesis [3, 5, 7, 8, 13]. In this respect, the disordered perovskites $La_{1-x}Bi_xMnO_3$ with FMI properties [14, 15] and multiferroics $La_{0.1}Bi_{0.9}MnO_3$ (Ref. [16]) and $La_{0.2}Bi_{0.8}MnO_3$ (Ref. [17]) are noteworthy to discuss. These La-substituted $BiMnO_3$ phases show multiferroic properties (Fig. 4.1) at low temperature (T $\approx$ 100 K) [16, 17]. For Bi-rich phase ($x \geq 0.6$), high-pressure/temperature synthesis is required and the multiferroicity is reported only for epitaxial thin films [16, 17]. However, the ambient pressure synthesis of Bi-centered ferromagnetic perovskites with similar characteristics (FMI) is highly desirable.

In this book, we have presented investigations on Bi-centered ferromagnetic perovskites, a prospective material for multiferroic application. Central focus is on $La_{0.5}Bi_{0.5}MnO_3$ and Bi-doped $La_2Mn(Co/Ni)O_6$ perovskites with various substitutions at different level. Hence, we have discussed the magnetotransport properties of the Bi-centered manganite perovskites where Mn^{3+} is partly replaced by Fe/Co/Ni-ions synthesized at atmospheric pressure and consequently demands to pay special attention for their ease of synthesis with desired properties (FMI). The phase purity of ambient pressure synthesized compounds is reflected from their X-ray

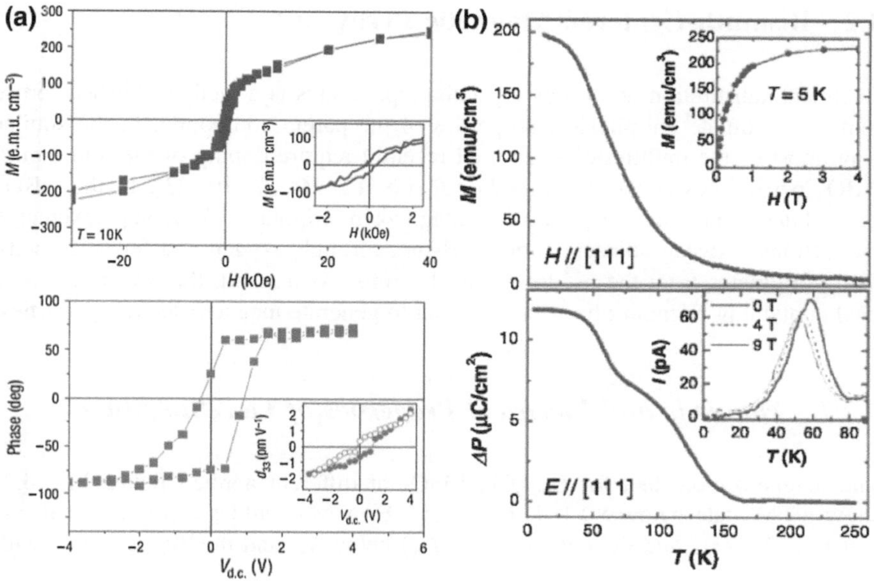

Fig. 4.1 Magnetic and ferroelectric response for **a** $La_{0.1}Bi_{0.9}MnO_3$ (adapted from Ref. [16]) and **b** $La_{0.2}Bi_{0.8}MnO_3$ (adapted from Ref. [17])

diffraction. Substitution of Bi for La in perovskite $LaMnO_3$ increases the insulating properties but decreases FM T_C [14, 15]. We have discussed various combinations of *A*- and *B*-site cations for different series of compounds, e.g., $La_{0.5}Bi_{0.5}Mn_{0.67}(Co/Ni)_{0.33}O_3$ (Refs. [18, 20, 21]), $La_{0.6}Bi_{0.4}Mn_{0.6}(Fe/Ni)_{0.4}O_3$ (Ref. [24]), $La_{1-x}Bi_xMn_{1-y}Fe_yO_3$ (Refs. [19, 23–25]), and $(La/Sr)_{1-x}Bi_xMn_{0.5}(Cr/Fe)_{0.5}O_3$ (Ref. [26]). Ferromagnetism with insulating/semiconducting properties has been discussed along with magnetodielectric effects. Dielectric properties have been highlighted for the doped phase $La_{0.5}Bi_{0.5}Mn_{0.67}Co_{0.33}O_3$ [21]. Additionally, colossal dielectric permittivities, capacitance with different thickness have been discussed for this class of semiconducting materials to distinguish them from the external factors such as interfacial polarization [21]. Finally, the results of high-quality epitaxial thin films of $La_{0.5}Bi_{0.5}Mn_{0.67}Co_{0.33}O_3$ have been discussed and also compared with the bulk phase [20]. In these systems the magnetization data reveal a strong magnetic anisotropy and a FM behavior with a superexchange interaction between transition metal cations, which are randomly distributed in the *B*-site. For some of the phases, the dielectric anomalies correspond to the onset of the magnetic ordering reflecting the coupling between the order parameters, i.e., the magnetoelectric coupling. The Raman spectroscopic studies supported the weak spin–lattice interaction around the magnetic transition, which explains the magnetodielectric properties at low temperature [20].

4.2 Bismuth-Centered Magnetic Perovskites

Chemical substitution at various crystallographic sites is a well-established fertile route to modulate the physical properties of the parent compound. In the similar way, most of the multiferroics reported recently require doping of the initial pure ABO_3 perovskites at the *A*-site and/or *B*-site [12, 16, 17, 19, 23]. In the ABO_3 perovskites containing bismuth, ferromagnetism originates from superexchange interactions between adjacent cations (*B*-site) through oxygen and ferroelectricity likely originates from the $6s^2$ lone pair electrons. As a result, the investigation of FMI containing bismuth offers the potential to generate magnetoelectric properties.

4.2.1 Magnetic and Electrical Properties of $La_{0.5}Bi_{0.5}MnO_3$

The magnetization data for $La_{0.5}Bi_{0.5}MnO_3$ at different applied fields (Fig. 4.2) show a FM state below 80 K [14, 15, 18]. However, unlike a conventional FM material, the ZFC and FC data diverge (T_{irr}) below T_C and the T_{irr} decreases with

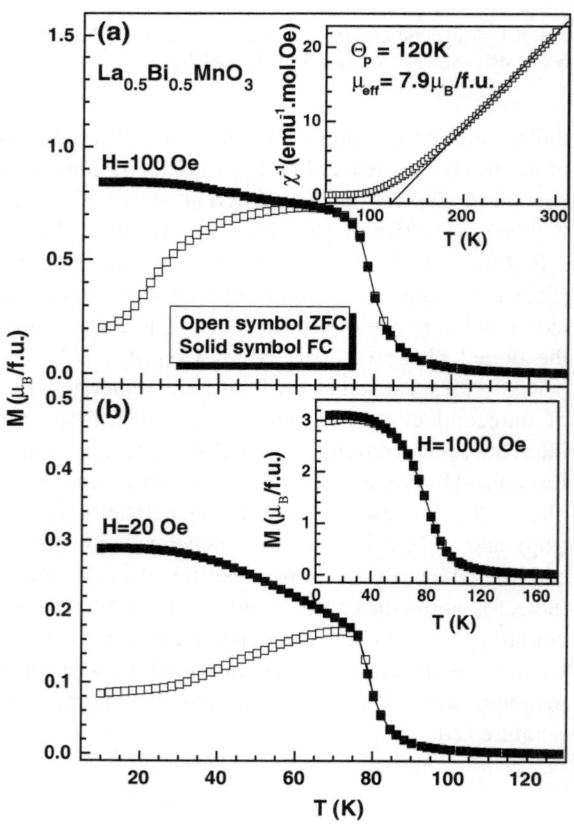

Fig. 4.2 Temperature variation of the ZFC (*open symbol*) and FC (*solid symbol*) magnetization, $M(T)$, of $La_{0.5}Bi_{0.5}MnO_3$ at different applied fields **a** H = 100 Oe (*inset* shows inverse susceptibility, χ^{-1}, vs. temperature plot) and **b** H = 20 Oe (*inset* figure for H = 1000 Oe) (adapted from Ref. [18])

the increase in field strength (Fig. 4.2). The nature of the FM phase is complex, and Zhao et al. [15] described as cluster glass, whereas weak ferromagnet or coexistence of paramagnetic (PM) and FM phases by Troyanchuk et al. [14]. The large T_{irr} at low fields is explained in the scenario of phase separation of FM domains involving Mn^{3+} and Mn^{4+} states distributed in an antiferromagnetic (AFM) matrix. Thus, during ZFC measurement the spins of magnetic ions freeze in random directions and the magnetic anisotropic energy are large. The low magnetic fields are not sufficient to align them in the direction of the applied field; as a result strong divergence appears [18]. But the anisotropic energy is overcome by higher field, and the spins are reoriented in the field direction, causing the superposition of ZFC and FC curves.

The merging of FC and ZFC curves at higher fields due to long-range FM ordering; however, the lack of magnetic saturation below the transition temperature and low field study point to the contrary. The isothermal magnetization, *M*(*H*), loops (Fig. 4.3a) confirm the short-range ordering, with the unsaturated behavior of *M*(*H*) loops even at higher fields, which is a characteristic feature of glassy-FM system [30, 31]. A distinct frequency-independent peak appeared in $\chi'(T)$ data, which corresponds to FM ordering (Fig. 4.3b). Below T_C, weak frequency-dependence behavior is noticed. The AC susceptibility behavior observed for $La_{0.5}Bi_{0.5}MnO_3$ is quite different from the canonical spin glass system [32–34] and is akin to that of glassy-FM materials [30, 31].

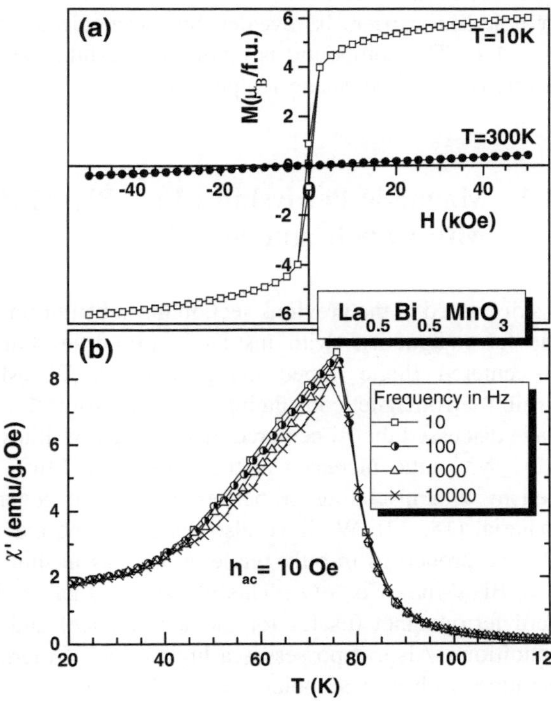

Fig. 4.3 **a** Field variation of magnetization at two different temperatures and **b** temperature variation in-phase component of AC susceptibility, χ', for $La_{0.5}Bi_{0.5}MnO_3$ (adapted from Ref. [18])

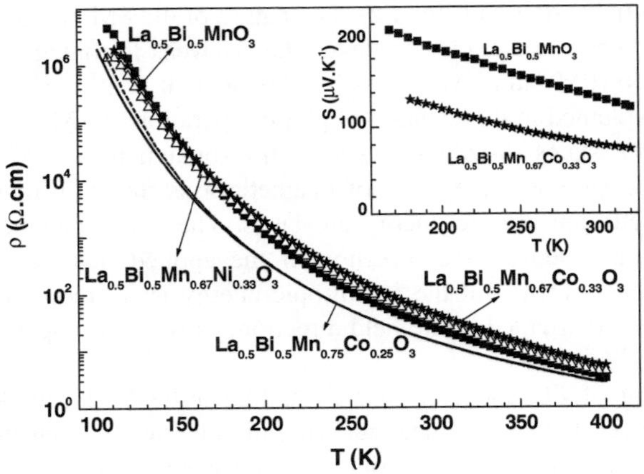

Fig. 4.4 Temperature-dependent electrical resistivity, $\rho(T)$, for $La_{0.5}Bi_{0.5}Mn_{1-x}(Co/Ni)_xO_3$. The *inset figure* shows Seebeck coefficient, $S(T)$, with the variation of temperature (adapted from Ref. [18])

Figure 4.4 represents the temperature-dependent electrical resistivity, $\rho(T)$, of $La_{0.5}Bi_{0.5}MnO_3$. With decreasing temperature the resistivity increases; magnitude is very high at low temperature. It does not exhibit any insulator–metal transition corresponding to FM T_C as it is commonly observed in conventional manganite perovskites. In other electron transport investigations, *p*-type polaronic conductivity or hole-like carriers is revealed by thermoelectric power measurements (inset in Fig. 4.4). The compound is reported to exhibit decrease of *p*-type polaronic conductivity with decreasing temperature.

4.3 Magnetic Perovskites $La_{0.5}Bi_{0.5}MnO_3$ Doped with Cobalt and Nickel

As discussed in the previous section the substitution of Bi for La in $LaMnO_3$ retains the ferromagnetism with insulating properties but decreases FM T_C. Likewise, Bi-centered *B*-site doped manganites $La_{0.5}Bi_{0.5}Mn_{1-x}M_xO_3$ with M = Co/Ni, exhibit ferromagnetic insulating properties for $0 \leq x \leq 0.33$ [18]. In this section, we have discussed the Bi-centered magnetic perovskites (La, Bi)(Mn, M)O_3, where the Mn^{4+}/M^{2+} ratio increases at the expense of Mn^{3+} in order to achieve ferromagnetism with insulating properties in view of generating multiferroics/spintronics material [18, 22]. We have also discussed the magnetoresistance and magnetodielectric properties in bulk phase as well as in thin films for a specific compound $La_{0.5}Bi_{0.5}Mn_{0.67}Co_{0.33}O_3$. This phase is attractive since at low temperature it is a hard ferromagnet (useful for memory device) and could be prepared at ambient conditions. Also, it possesses a high value of thermoelectric power at room temperature with *p*-type polaronic conductivity [18]. We have also extended our discussion for Ni-substituted phase, where ferromagnetism is enhanced at a lesser degree, with a T_C of 97 K for the compound $La_{0.5}Bi_{0.5}Mn_{0.67}Ni_{0.33}O_3$ [18].

4.3.1 Magnetotransport Properties of Single-Phase Bulk $La_{0.5}Bi_{0.5}Mn_{1-x}M_xO_3$ (M = Co, Ni)

It is reported that Co- and Ni substitution induces ferromagnetism in $LaMnO_3$ (Refs. [35–37]) and formulated ordered perovskites La_2MnCoO_6 (Refs. [26, 27, 38]), La_2MnNiO_6 (Refs. [9, 39]), and Bi_2MnNiO_6 [13, 28, 40]. We have discussed the effect of Co- and Ni substitution in $La_{0.5}Bi_{0.5}MnO_3$. The perovskite $La_{0.5}Bi_{0.5}Mn_{1-x}M_xO_3$, with M = Co/Ni, is reported as single phase over a small range of substitution, $0 \leq x \leq 0.33$, without any traces of impurities [18].

All these Bi-centered perovskites are reported to crystallize in orthorhombic structure, with the *Pnma* space group (disordered perovskite). The substitution of cobalt for manganese in perovskite $La_{0.5}Bi_{0.5}MnO_3$, significantly increases the FM T_C [18]. Figure 4.5 shows the ZFC and FC magnetization for $La_{0.5}Bi_{0.5}Mn_{1-x}Co_xO_3$ series, the highest $T_C \approx 130$ K, for $x = 0.33$. The increasing trend of FM interactions with higher cobalt doping was explained by superexchange interactions due to the presence of Co^{2+} and Mn^{4+} ions [26, 27, 38]. High-field magnetization data for Co-substituted samples is shown in Fig. 4.6.

Unlike $La_{0.5}Bi_{0.5}MnO_3$, the Co-substituted samples show considerably large divergence between ZFC and FC magnetization at low temperatures even at high field. The cusp in ZFC data becomes broad at higher field and shifts toward lower temperatures. The *M(H)* loop (insets of Fig. 4.6), depicts unsaturated values of

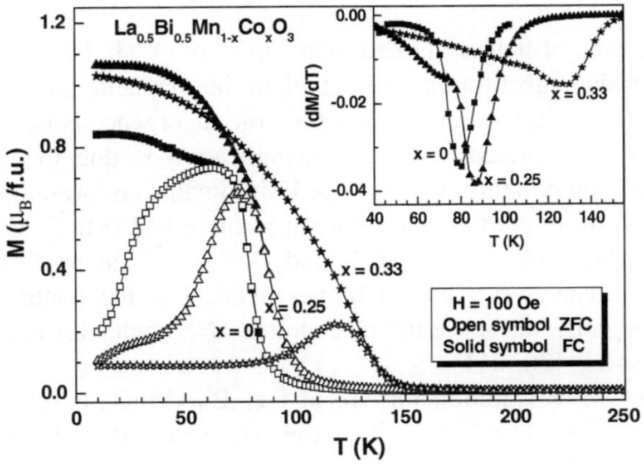

Fig. 4.5 Temperature-dependent ZFC (*open symbol*) and FC (*solid symbol*) magnetization, *M(T)*, for $La_{0.5}Bi_{0.5}Mn_{1-x}Co_xO_3$. The *inset* shows (d*M*/d*T*) versus temperature plot for FC magnetization (adapted from Ref. [18])

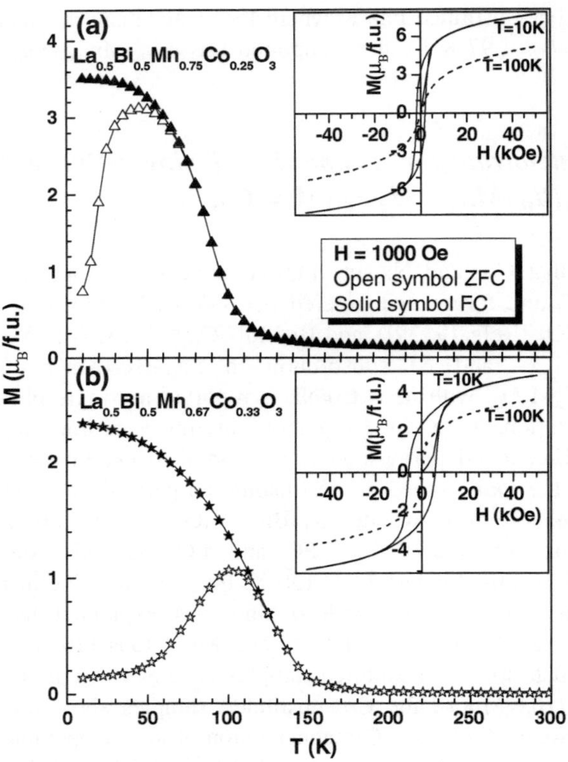

Fig. 4.6 Temperature-dependent ZFC (*open symbol*) and FC (*solid symbol*) magnetization, *M*(*T*), for **a** $La_{0.5}Bi_{0.5}Mn_{0.75}Co_{0.25}O_3$ and **b** $La_{0.5}Bi_{0.5}Mn_{0.67}Co_{0.33}O_3$. The *insets* show typical hysteresis curves at two different temperatures (adapted from Ref. [18])

magnetization, even at higher applied field (up to 5 T). Higher values of coercive field with increasing substitution levels explain the magnetic anisotropy below the FM transition. This might be the origin of a large divergence between the ZFC and FC curves. The unsaturated value of magnetization is due to electronic phase separation at low temperature, where large FM domains are present inside an AFM matrix. There is strong competition between positive FM (Mn^{3+}–Mn^{4+} and Mn^{4+}–Co^{2+}) and negative AFM (Mn^{4+}–Mn^{4+} and Co^{2+}–Co^{2+}) interactions and the FM interactions dominate over AFM at higher fields. But the contribution of AFM interactions is significant; hence the unsaturated *M*(*H*) behavior is reported akin to glassy-FM materials [30, 31].

The effect of Ni substitution for Mn in $La_{0.5}Bi_{0.5}MnO_3$ is quite similar to Co substitution (Fig. 4.7). The FM T_C increases to a value of 97 K for $x = 0.33$, but lower than Co-substituted phase. The field variation of magnetization provides a lower value of magnetic moment for Ni phases with a small coercive field, 30 Oe at 10 K and magnetic moment of 3.5 μ_B/f.u [18]. This is the lowest reported value of

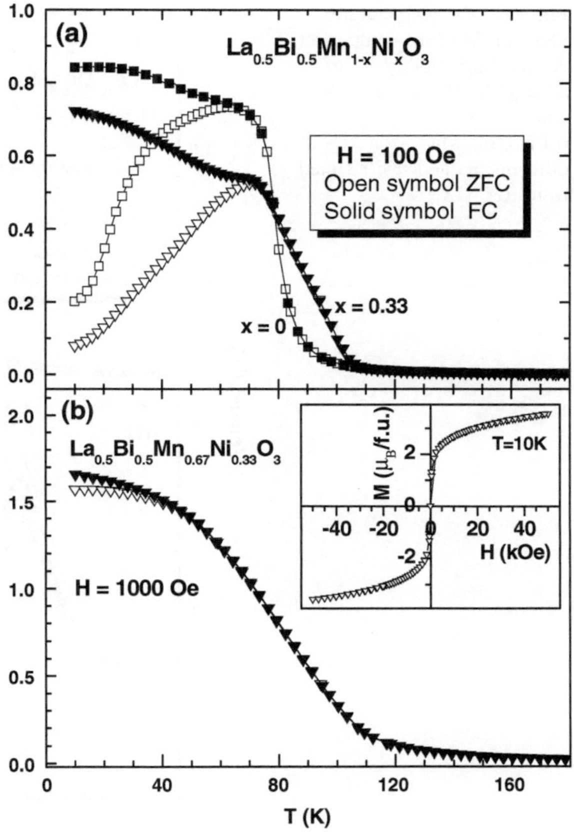

Fig. 4.7 Temperature variation of the ZFC (*open symbol*) and FC (*solid symbol*) magnetization, *M*(*T*), of **a** $La_{0.5}Bi_{0.5}Mn_{1-x}Ni_xO_3$ (H = 100 Oe) and **b** $La_{0.5}Bi_{0.5}Mn_{0.67}Ni_{0.33}O_3$ (H = 1000 Oe); the *inset* shows typical hysteresis curve at 10 K (adapted from Ref. [18])

moment, although the magnetic interactions are similar in nature, i.e., FM interactions between Mn^{3+}–Mn^{4+} and Mn^{4+}–Ni^{2+} [9, 39]. The smaller value of moment and lower value of T_C (≅97 K), compared to the cobalt phase is explained by its non-stoichiometric nature, which induces a smaller Mn^{4+} content and creates disorder on the cationic sites [18].

Figure 4.8 shows AC susceptibility for $La_{0.5}Bi_{0.5}Mn_{0.67}(Co/Ni)_{0.33}O_3$. The doped phases also follow the low-field ZFC magnetization. The Co-substituted phase reveals a weak frequency-dependent peak at low temperature, corresponding to the FM ordering. However, the Ni-substituted phase does not show any shift in the peak temperature with varying frequencies.

Hence, the Co-substituted phase has a frequency-dependent maximum in χ' (Fig. 4.8a), while the Ni-substituted phase (Fig. 4.8b) reveals a similar feature to parent $La_{0.5}Bi_{0.5}MnO_3$ (Fig. 4.3b). The magnetic behavior of $La_{0.5}Bi_{0.5}Mn_{0.67}Co_{0.33}O_3$ is consistent with the materials behaving as a spin-glass-like system [32–34]. In contrast, the parent $La_{0.5}Bi_{0.5}MnO_3$ and Ni-substituted phases show a glassy-FM behavior [30, 31].

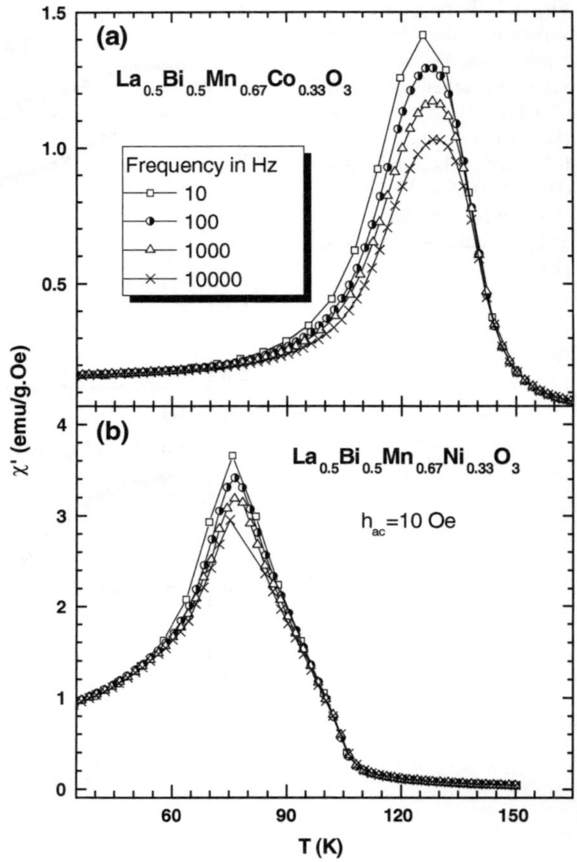

Fig. 4.8 The temperature variation in-phase component of AC susceptibility, χ', of **a** $La_{0.5}Bi_{0.5}Mn_{0.67}Co_{0.33}O_3$ and **b** $La_{0.5}Bi_{0.5}Mn_{0.67}Ni_{0.33}O_3$ at different frequencies (adapted from Ref. [18])

The $\rho(T)$ of $La_{0.5}Bi_{0.5}Mn_{1-x}(Co/Ni)_xO_3$ series is shown in Fig. 4.4. With decreasing temperature, the resistivity increases and the value is very high at low temperature (≤100 K) [18]. The rapid change in temperature coefficient of resistivity ($d\rho/dT$) from room temperature to low temperatures signifies the insulating behavior. The magnetoresistance effect in an applied field of 7 T is very small at low temperature. In the 100–400 K range, the temperature variation of resistivity (with magnetic field of 0 and 7 T) confirms the insulating phase, although the materials are FM below room temperature, as revealed from the magnetization results discussed earlier. Thermoelectric power measurements give an idea about the type of charge carriers, and the substitution level depending upon carrier concentration in the materials. The Seebeck coefficient value (inset Fig. 4.4) is reported around +133 μV/K for $La_{0.5}Bi_{0.5}MnO_3$ at room temperature, which gradually increases with decrease in temperature. For doped phase $La_{0.5}Bi_{0.5}Mn_{0.67}Co_{0.33}O_3$, the value ≈81 μV/K at room temperature, which confirms the *p*-type polaronic conductivity or hole-like carriers in the materials, similar to the perovskite manganites [35–37].

4.3.2 Dielectric Properties of $La_{0.5}Bi_{0.5}Mn_{0.67}Co_{0.33}O_3$

4.3.2.1 Bulk Phase of $La_{0.5}Bi_{0.5}Mn_{0.67}Co_{0.33}O_3$

As discussed in the preceding section that the single-phase perovskite $La_{0.5}Bi_{0.5}Mn_{0.67}Co_{0.33}O_3$ exhibits semiconducting-type behavior. Interestingly, this phase shows a high value of thermoelectric power at room temperature [18]. In this section we will discuss the dielectric behavior and extend to similar type of semiconducting phases, which might result from artifacts owing to the charge injection at the electrode contacts [41, 42]. This type of feature is not uncommon for the semiconducting samples compared to insulators, which is always present in biological samples [43] and increase the dielectric response toward colossal values (especially at lower frequencies). Thus, the high dielectric value is indeed controversial. Disagreement exists on this, and the anomalous dielectric responses have been assigned to electrodes polarization [41] or grain boundary artifacts [44]. In addition, a colossal dielectric constant with a large magnetocapacitance value is observed frequently in bulk semiconducting samples where different cations are responsible for their magnetic and dielectric responses [45–49].

In this section, we have discussed the complex impedance ($Z = Z' + iZ''$) of a bulk $La_{0.5}Bi_{0.5}Mn_{0.67}Co_{0.33}O_3$ sample of variable thicknesses (D), as a function of temperature and frequency. Filippi et al. [21] have analyzed capacitance (C) as a function of reciprocal thickness, D^{-1}, for true bulk response (non-interfacial). Figure 4.9 illustrates the dielectric permittivity (ε) and dielectric loss ($\tan\delta$) as a function of temperature [21]. At lower temperatures, ε exhibits a small value $\sim$40, with an initial sharp upturn (accompanied by a peak in $\tan\delta$) around 120 K, reaching values higher than 1000. On further increase in temperature, ε increases and attains a higher value of 10^4. All these features are commonly observed in bulk semiconducting samples [41, 42, 47, 48].

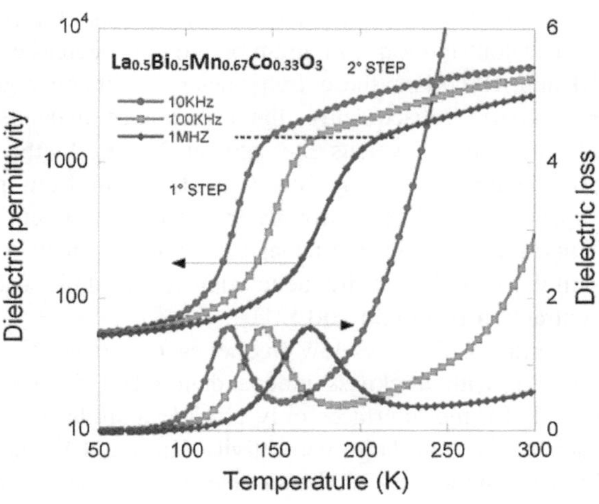

Fig. 4.9 Temperature dependence of dielectric permittivity and dielectric loss at certain frequencies for $La_{0.5}Bi_{0.5}Mn_{0.67}Co_{0.33}O_3$ (adapted from Ref. [21])

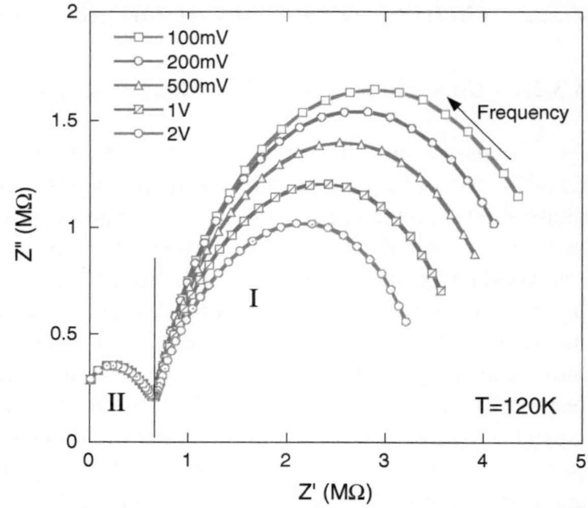

Fig. 4.10 Impedance complex plane Nyquist plot for $La_{0.5}Bi_{0.5}Mn_{0.67}Co_{0.33}O_3$ at 120 K for various AC voltage amplitude between 0.1 and 2 V (adapted from Ref. [21])

At low temperature (10 K), the DC resistivity exceeds 10^7 Ω cm, and the $La_{0.5}Bi_{0.5}Mn_{0.67}Co_{0.33}O_3$ behaves as a pure capacitor (i.e., the current–voltage phase shift is almost 90°). With increasing temperature a second contribution appears because of low resistivity which results a low frequency arc in the Nyquist plot (Fig. 4.10). The Z'' versus Z' Nyquist plot at 120 K, shows a low frequency dominating arc (right side Fig. 4.10), with strong nonohmic response to the AC voltage. The second arc at higher frequency is not sensitive to the voltage amplitude, and represents the intrinsic sample response [21].

Figure 4.11 shows a linear dependency of C with D^{-1} at different temperatures. The capacitance for each frequency is linear in D^{-1} and extrapolates to zero (at 10 K), confirming the intrinsic response (Fig. 4.11a). Interface effect appears at high temperatures ($T > 120$ K), which is manifested in the form of large offset in the C versus D^{-1} plots (Fig. 4.11b, c) and is also linear at higher frequencies.

The contribution from grain boundaries decreases with frequency, and persists at 10 and 100 kHz (these frequencies lie in the region between the two arcs of Fig. 4.10). Consequently, the capacitance depends on frequency. Figure 4.11c illustrates the C versus D^{-1} plot at 200 K. In this region DC resistivity is low (10^3 Ω cm) and a significant contribution from polarization of the electrodes appears. The offset is strongly frequency-dependent and larger at low frequencies. Thus, the parasitic interfacial capacitance, which is physically in series with the sample, results in a frequency-dependent shift. The calculated permittivity after subtraction is around 900 [21].

Figure 4.12 shows low frequency C versus D^{-1} plot for indium electrodes (at 200 K), with thickness independent offset for low frequency. At 300 K (inset Fig. 4.12), the interfaces only provide a slight contribution to the high-frequency capacitance. The large permittivity (around 10^4) measured at 300 K is due to the higher conductivity of the sample. Unlike the previously reported results on this

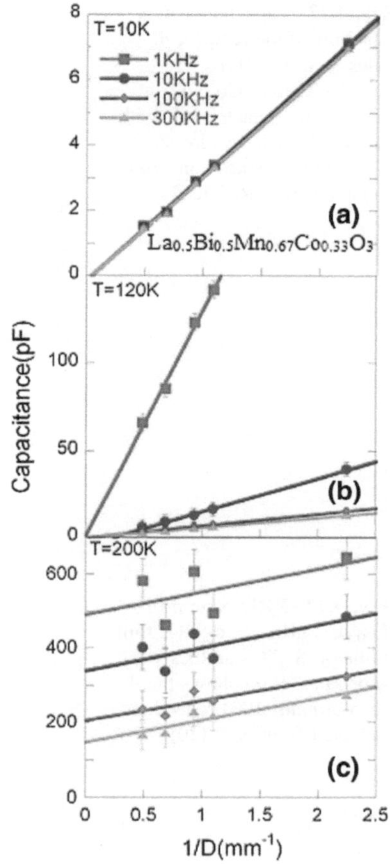

Fig. 4.11 Capacitance as a function of the reciprocal thickness (D^{-1}) for $La_{0.5}Bi_{0.5}Mn_{0.67}Co_{0.33}O_3$ at **a** 10 K, **b** 120 K, and **c** 200 K (adapted from Ref. [21])

field (origin of giant permittivity and artifacts coming from conductivity), Filippi et al. [21] derived a model from experimental data, on the basis of its variation with frequency, voltage, and sample thickness. The results indicate an interfacial contribution (not *Maxwell–Wagner*-like) modeled as a capacitor in parallel with the sample, giving a monotonic contribution as a function of the temperature.

4.3.2.2 Thin Films of $La_{0.5}Bi_{0.5}Mn_{0.67}Co_{0.33}O_3$

In this section, we have discussed magnetic, dielectric, and magnetodielectric behaviors of $La_{0.5}Bi_{0.5}Mn_{0.67}Co_{0.33}O_3$ epitaxial thin films [20]. Figure 4.13 shows the XRD pattern of $La_{0.5}Bi_{0.5}Mn_{0.67}Co_{0.33}O_3$ on $SrTiO_3$ (STO) and $LaAlO_3$ (LAO) substrates revealing the epitaxial nature of the film.

Figure 4.14 shows the temperature-dependent magnetization of the $La_{0.5}Bi_{0.5}Mn_{0.67}Co_{0.33}O_3$ film on LAO substrate at two configurations. First, when

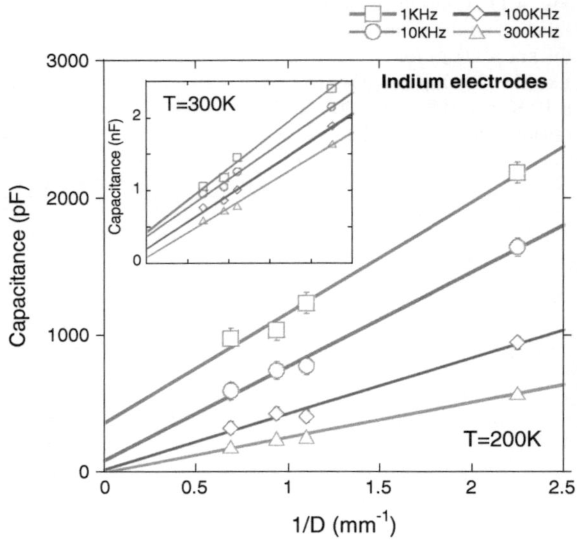

Fig. 4.12 Capacitance as a function of the reciprocal thickness (1/*D*) for $La_{0.5}Bi_{0.5}Mn_{0.67}Co_{0.33}O_3$ at 200 K, for contacts made with indium. The same plot at *T* = 300 K is shown in *inset* (adapted from Ref. [21])

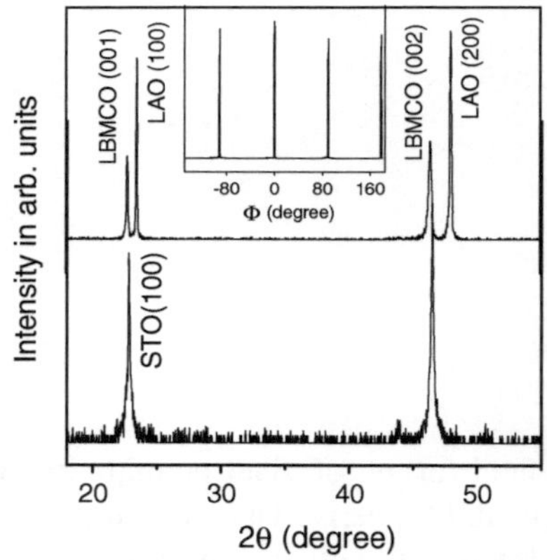

Fig. 4.13 XRD pattern of $La_{0.5}Bi_{0.5}Mn_{0.67}Co_{0.33}O_3$ thin films on $SrTiO_3$ and $LaAlO_3$ substrates. *Inset* shows the Φ-scan around (103) reflection (adapted from Ref. [20])

the applied magnetic field is parallel $[H\|(100)_S]$ to the substrate surface, and second, when the applied magnetic field is perpendicular $[H\|(001)_S]$. The highest magnetization (2.21 μ_B/f.u) has been reported for the magnetic field parallel to substrate (film) surface $[H\|(100)_S]$ in comparison to the perpendicular field $[H\|(001)_S]$. A strong magnetic anisotropy (along the directions parallel and perpendicular of the film plane) has been reported, which is associated with a single domain or epitaxial orientation of the films.

The inset of Fig. 4.14 shows the hysteresis loop for the films at 10 and 100 K. The *M*(*H*) loops and the strong divergence T_{irr} at low temperature are similar to bulk phase, the H_C values for the thin films are greater than the bulk phase [18]. The T_{irr} is higher for LAO film (T_{irr} ~ 100 K) compared to STO film (85 K). Low

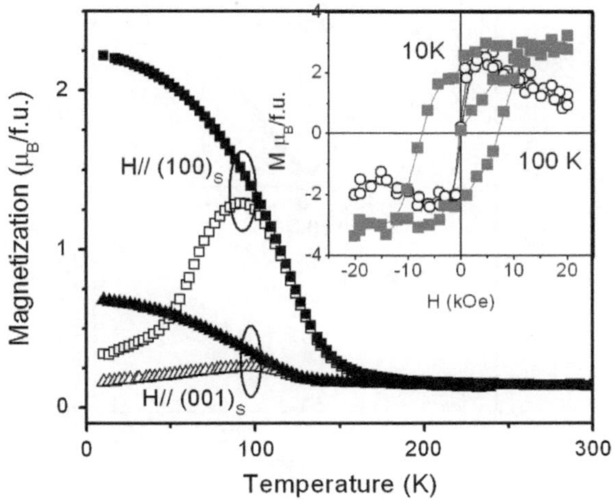

Fig. 4.14 Temperature-dependent ZFC (*open symbol*) and FC (*solid symbol*) magnetization *M(T)*, of $La_{0.5}Bi_{0.5}Mn_{0.67}Co_{0.33}O_3$ thin film on $LaAlO_3$ (001) substrates (H = 1000 Oe, applied parallel to film surface $H||(100)_S$, and perpendicular to the film surface, $H||(001)_S$. The *insets* show the magnetic hysteresis curves at different temperatures ($H||(100)_S$) (adapted from Ref. [20])

temperature magnetic properties prove the existence of ferromagnetism in film, and the existence of FM interaction between Mn^{4+} and Co^{2+} cations via superexchange mechanism [26–28].

Figure 4.15 shows the temperature-dependent dielectric constant for thin film and bulk phase. In contrast to the film, the bulk phase exhibits a rapid increase in the dielectric constant above 150 K (Fig. 4.15), due to an extrinsic Maxwell–Wagner kind relaxation [49, 50]. The dielectric constant exhibits an anomaly in both the thin film and the bulk phase at around 200 K, with frequency-dependent behavior as expected from extrinsic effect [49, 50]. An additional dielectric anomaly is observed only for thin films around the magnetic transition (123 K). This is due to the presence of a weak spin lattice interaction in the thin film around the magnetic transition. The onset of interactive ordered magnetic clusters gives rise to spin lattice coupling, which distorts the lattice, and alters the dielectric constant [51]. The presence of lone pair electrons of bismuth facilitates lattice distortion at the onset of magnetic ordering. The dielectric anomaly around the magnetic transition is dominated by the extrinsic Maxwell–Wagner effect at low frequencies (<500 kHz). A positive magnetodielectric effect $\Delta\varepsilon$ of 0.7 % is reported at 10 K (inset Fig. 4.15) [20]. This small effect has been reported up to 100 K, which is close to the magnetic transition, without the overlap of other extrinsic effects and attributed to the intrinsic spin–lattice coupling [52].

Figure 4.16 shows the polarized Raman spectra of $La_{0.5}Bi_{0.5}Mn_{0.67}Co_{0.33}O_3$ thin films. The Raman active modes exhibit a very large full width at half maxima value. Broadness in the phonon peaks was suggested to be associated with random distribution of Mn^{4+} and Co^{2+} ions. Moreover, the apparent lack of polarization

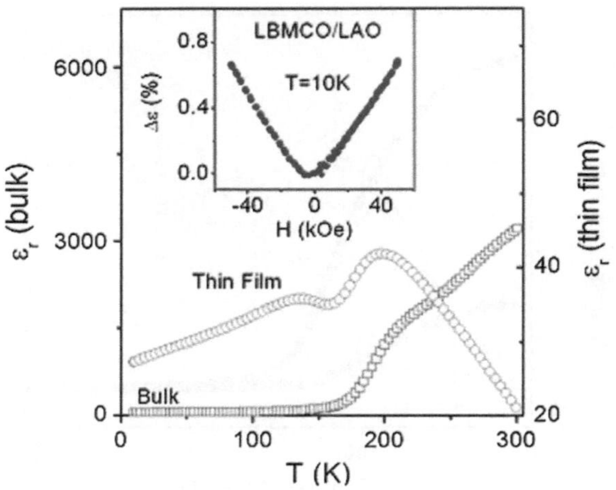

Fig. 4.15 Temperature-dependent relative permittivity (ε_r) for $La_{0.5}Bi_{0.5}Mn_{0.67}Co_{0.33}O_3$ compound. The *inset* shows the variation of $\Delta\varepsilon$ % $\{[\varepsilon(H) - \varepsilon(0)/\varepsilon(0)] \times 100\}$ with magnetic field at 10 K for thin film (adapted from Ref. [20])

Fig. 4.16 Raman spectrum of $La_{0.5}Bi_{0.5}Mn_{0.67}Co_{0.33}O_3$/$LaAlO_3$ film. *Inset* shows the softening of a Raman mode with temperature in both substrates (adapted from Ref. [20])

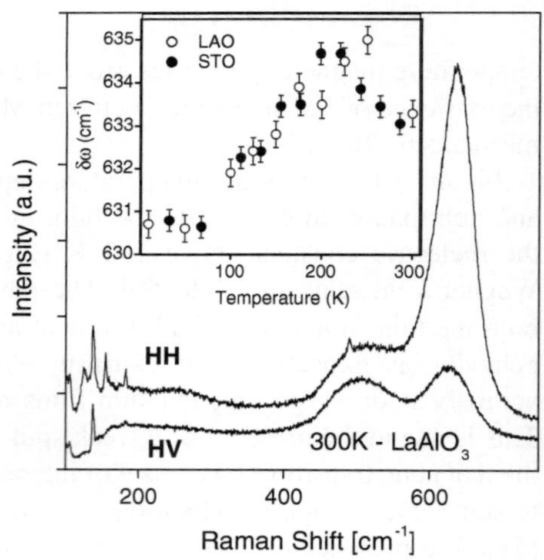

dependence further attests to the lack of a cation ordering. Temperature-dependent shift of the 633 cm^{-1} Raman mode shows the spin–lattice interactions (inset of Fig. 4.16). The modulation of the superexchange integral by the phonons is generally observed as softening of phonons [53], which occurs in the vicinity of the magnetic transition temperature. The magnetic transition temperature is about 123 K for thin film (Fig. 4.14), which is much lower than the observed phonon peak shifts at 230 K (Fig. 4.16). This behavior correlates well with the temperature dependence of the dielectric properties of these films, which also peak around the same temperature. The same ambiguity has been also reported in other magnetodielectric materials, such as $DyMnO_3$ [54].

4.4 Other Bismuth-Centered Magnetic Perovskites

4.4.1 *Magnetotransport Properties of Bulk $La_{0.6}Bi_{0.4}Mn_{1-x}(Co/Fe/Ni)_xO_3$*

Single-phase bulk perovskites of the type $La_{0.6}Bi_{0.4}Mn_{1-x}(Co/Fe/Ni)_xO_3$, are discussed for $x = 0.0–0.4$, with orthorhombic structure (*Pnma* space group) similar to previous perovskites [18–24]. The magnetic and electrical properties of these perovskites are stimulating an extensive investigation because of their novel functionalities and applications.

Single-Phase Bulk $La_{0.6}Bi_{0.4}MnO_3$ and $La_{0.6}Bi_{0.4}Mn_{0.6}Co_{0.4}O_3$

The parent perovskite $La_{0.6}Bi_{0.4}MnO_3$ shows a PM to FM transition, T_C, at 103 K, whereas the substituted phase $La_{0.6}Bi_{0.4}Mn_{0.6}Co_{0.4}O_3$ has a high value of FM $T_C \sim 181$ K (Fig. 4.17a) [22]. Moreover, there is a considerable divergence for Co-substituted phase (for 100 and 5000 Oe), whereas the parent phase $La_{0.6}Bi_{0.4}MnO_3$ shows weak irreversibility at low magnetic field (≤100 Oe), akin to $La_{0.5}Bi_{0.5}(Mn,Co/Ni)O_3$ perovskites [18]. The magnetic interactions in all the Bi-centered perovskites are different from the long-range ferromagnetism. The *M* (*H*) loop for $La_{0.6}Bi_{0.4}MnO_3$ is very small, whereas for $La_{0.6}Bi_{0.4}Mn_{0.6}Co_{0.4}O_3$ are relatively large with a higher coercive field (inset of Fig. 4.17a). The *M*(*H*) loops do not saturate even at higher fields, which is a characteristic feature of a glassy ferromagnets [30, 31]. At low temperatures, there is subtle balance between the FM and AFM interactions or in other words the system is electronically phase separated into FM and AFM clusters. A competition between them gives rise to glassy FM state [30, 31].

Single-Phase Bulk $La_{0.6}Bi_{0.4}Mn_{0.6}(Fe/Ni)_{0.4}O_3$

A comprehensive discussion on the effect of Fe- and Ni substitution for Mn in the parent bismuth perovskite $La_{0.6}Bi_{0.4}MnO_3$ is presented here. Both the substituted phases show an impressive change in magnetic ordering temperature [24]. The shifts in T_C of these doped phases are in opposite direction with respect to the parent $La_{0.6}Bi_{0.4}MnO_3$ phase. Figure 4.18 shows the ZFC and FC magnetization for $La_{0.6}Bi_{0.4}Mn_{0.6}(Fe/Ni)_{0.4}O_3$. The Ni-substituted phase shows an increase in $T_C \sim 200$ K, whereas for Fe-substituted phase T_C decrease to 95 K. With decreasing the temperature, the *M*(*T*) for Fe-substituted phase shows a FM to AFM-like transition around 60 K (Fig. 4.18a), where the ZFC magnetization decreases rapidly to a lower value. The ZFC and FC magnetization curves illustrate a large divergence for both the samples, for a lower applied field (≤100 Oe), akin to

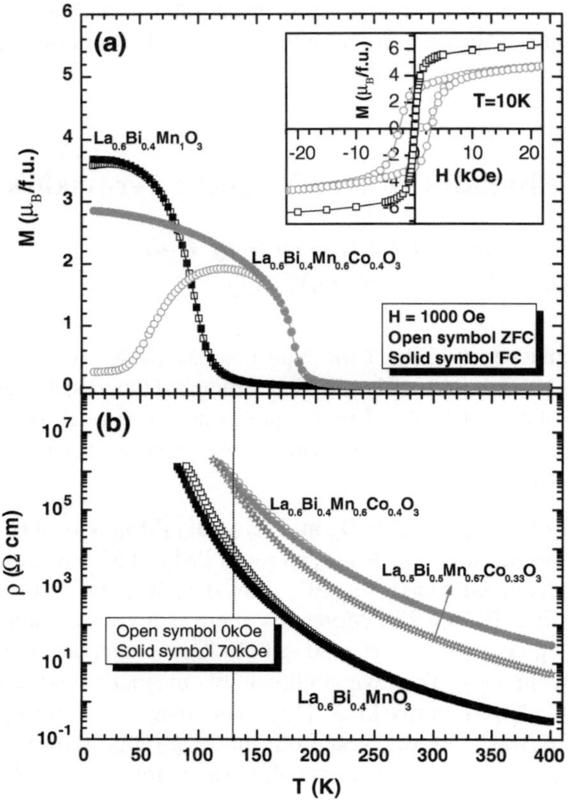

Fig. 4.17 **a** Temperature-dependent ZFC (*open symbol*) and FC (*solid symbol*) magnetization, *M* (*T*), *Inset figure* shows the, *M*(*H*), loops at 10 K; and **b** Electrical resistivity, $\rho(T)$, for $La_{0.6}Bi_{0.4}MnO_3$ and $La_{0.6}Bi_{0.4}Mn_{0.6}Co_{0.4}O_3$ (adapted from Ref. [22])

the earlier compounds [18–24]. The magnetic interactions in Fe/Ni-substituted samples are due to short-range type ferromagnetism. For Ni phase, the $T_{irr} \sim 179$ K and two broad magnetic transitions appear at 90 and 170 K (Fig. 4.18b). With higher field value, T_{irr} shifts toward lower temperature and the curves merge. In contrast, for $La_{0.6}Bi_{0.4}Mn_{0.6}Fe_{0.4}O_3$, curves merge above the AFM-like transition ($T_{irr} \sim 60$ K) and the ZFC cusp remains prominent (Fig. 4.18a), similar to spin or cluster glass type material [30–34].

The *M*(*H*) loops for the Fe- and Ni-substituted phases show a distinctive feature (Fig. 4.19). A narrower hysteresis loop is observed for $La_{0.6}Bi_{0.4}Mn_{0.6}Ni_{0.4}O_3$, whereas for $La_{0.6}Bi_{0.4}Mn_{0.6}Fe_{0.4}O_3$ loops are relatively larger [24]. Moreover, for $La_{0.6}Bi_{0.4}Mn_{0.6}Fe_{0.4}O_3$, the S-shaped hysteresis curves (Fig. 4.19a) are noticed below T_C akin to glassy FM system [30, 31]. The loop originates from the superposition of two types of contributions below the FM transitions. The FM component is characterized by a hysteresis loop and a finite value of coercive field

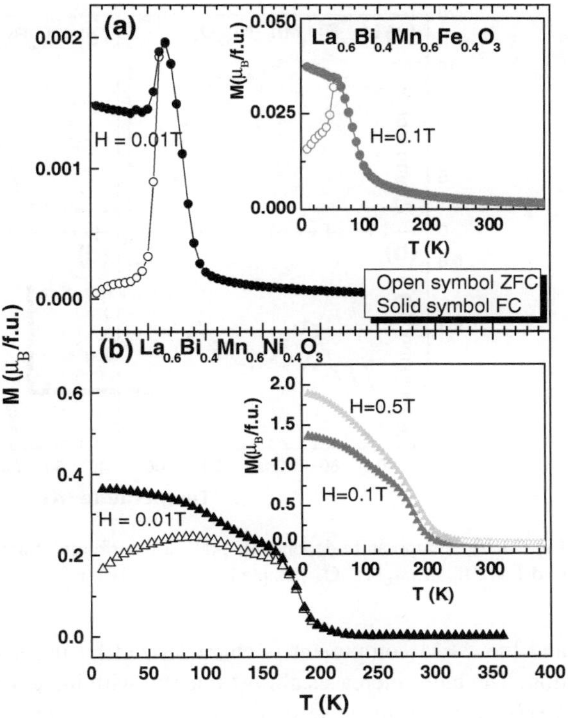

Fig. 4.18 Temperature-dependent ZFC (*open symbol*) and FC (*solid symbol*) magnetization, M(T), for **a** $La_{0.6}Bi_{0.4}Mn_{0.6}Fe_{0.4}O_3$ and **b** $La_{0.6}Bi_{0.4}Mn_{0.6}Ni_{0.4}O_3$ in different applied fields (H = 0.01, 0.1 and 0.5 T) (adapted from Ref. [24])

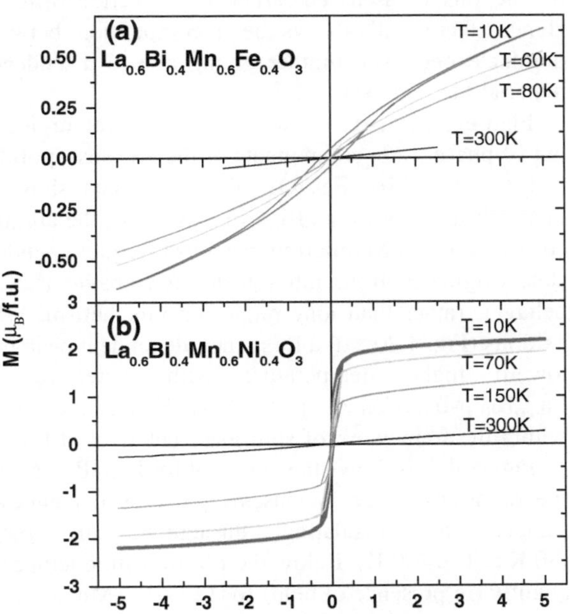

Fig. 4.19 The *M*(*H*) loops at different temperatures for **a** $La_{0.6}Bi_{0.4}Mn_{0.6}Fe_{0.4}O_3$ and **b** $La_{0.6}Bi_{0.4}Mn_{0.6}Ni_{0.4}O_3$ (adapted from Ref. [24])

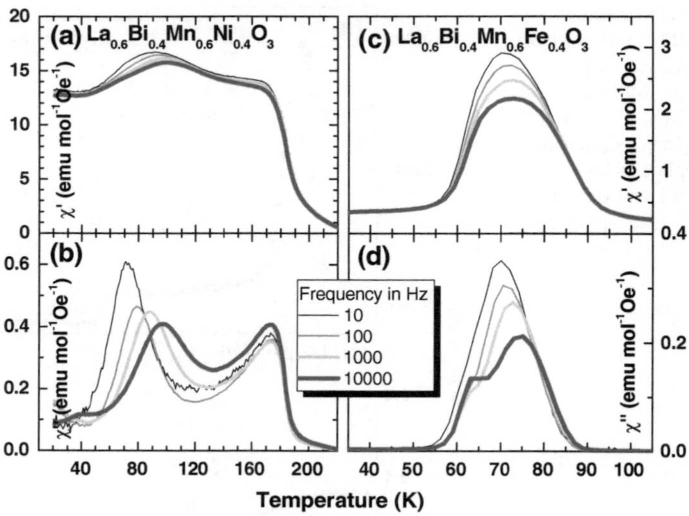

Fig. 4.20 Temperature-dependent AC susceptibility for **a**, **b** $La_{0.6}Bi_{0.4}Mn_{0.6}Ni_{0.4}O_3$ and **c**, **d** $La_{0.6}Bi_{0.4}Mn_{0.6}Fe_{0.4}O_3$ (adapted from Ref. [24])

and the AFM component is characterized by unsaturated value of the magnetization. The latter increases almost linearly with increasing applied field. The observed $M(H)$ behavior in these compounds is due to the presence of different magnetic interactions as discussed earlier in connection with the electronic phase separation. Hence, there will always be a competition between these two interactions to dominate one over another giving rise to a tendency of glassy FM state in the material [18, 30, 31].

Figure 4.20a–d shows the temperature-dependent in-phase $\chi'(T)$ and out-of-phase $\chi''(T)$ components of the AC susceptibility for $La_{0.6}Bi_{0.4}Mn_{0.6}Ni_{0.4}O_3$ and $La_{0.6}Bi_{0.4}Mn_{0.6}Fe_{0.4}O_3$. The $\chi'iT)$ data show similar features to the ZFC magnetization. For $La_{0.6}Bi_{0.4}Mn_{0.6}Ni_{0.4}O_3$, the abrupt increase of $\chi'(T)$ at ~ 190 K corresponds to FM ordering, is almost frequency-independent (Fig. 4.20a). The FM state originates from intra-cluster ferromagnetism, i.e., from cluster glass type behavior rather than long-range ferromagnetism. Whereas $La_{0.6}Bi_{0.4}Mn_{0.6}Fe_{0.4}O_3$ exhibits (Fig. 4.20c, d) a frequency-dependent peak below T_C (~ 70 K) which shifts toward higher temperature with increasing frequency similar to the $La_{0.6}Bi_{0.4}Mn_{0.6}Co_{0.4}O_3$ phase [24]. Hence, the Fe-substituted phase is explained within the framework of spin glass behavior at low temperatures.

Figure 4.17b shows the $\rho(T)$ plot for $La_{0.6}Bi_{0.4}MnO_3$, in the zero-field condition, the resistivity value increases rapidly as manganese is replaced by cobalt. The samples are insulating throughout the measured temperature range ($80 \text{ K} \le T \le 400$ K). Below the FM transition temperature, the $\rho(T)$ value decreases slightly (in presence of field) for $La_{0.6}Bi_{0.4}Mn_{0.6}Co_{0.4}O_3$ and the effect is larger for the parent $La_{0.6}Bi_{0.4}MnO_3$ compound. However, none of these phases show an

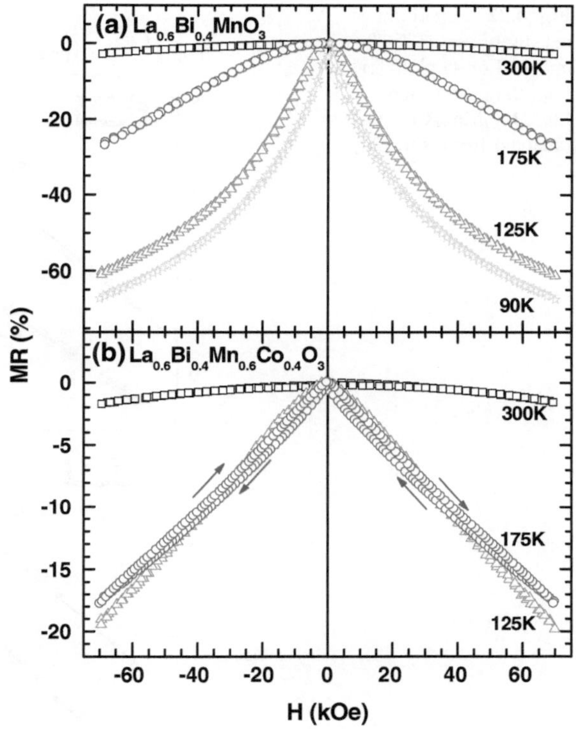

Fig. 4.21 Isothermal magnetoresistance MR(%) at different temperatures for **a** $La_{0.6}Bi_{0.4}MnO_3$ and **b** $La_{0.6}Bi_{0.4}Mn_{0.6}Co_{0.4}O_3$ (adapted from Ref. [22])

insulator–metal transition, even in the presence of higher magnetic field. In the 80–400 K range, the $\rho(T)$ curve confirms the insulating behavior (Fig. 4.17b), although both of them being FM below room temperature and the MR effect is significant. Hence, at low temperature there is a strong correlation between magnetic and electronic phase in these systems.

For the parent $La_{0.6}Bi_{0.4}MnO_3$ phase, the highest MR value is around −67 % at 90 K (Fig. 4.21a) and near T_C the value is ~−61 %. Similarly, for $La_{0.6}Bi_{0.4}Mn_{0.6}Co_{0.4}O_3$ the MR value is −17 %, just below the T_C (at 175 K) and the highest value is ~−19 % at 125 K (Fig. 4.21b). The occurrence of anisotropic MR behavior at low temperatures similar to those in $M(H)$ studies suggests the strongly correlated nature of field-induced magnetic and electronic behavior. Moreover, for $La_{0.6}Bi_{0.4}MnO_3$ the MR value in the PM region (Fig. 4.21a) is −25 % (at 175 K), which is due to inter-grain MR effect above T_C [55]. Likewise, for $La_{0.6}Bi_{0.4}Mn_{0.6}Fe_{0.4}O_3$ and $La_{0.6}Bi_{0.4}Mn_{0.6}Ni_{0.4}O_3$, the isothermal MR values (at 125 K) are around −2.5 and −20 %, respectively [24].

It is well known that in compounds with competing magnetic orders, a magnetic field favoring one kind of order in the spins also causes a large negative MR. Therefore, field induced magnetotransport interactions are expected due to the presence of magnetic clusters. The negative MR for the compounds is understood in the scenario of suppression of the electron scattering below the FM T_C in the

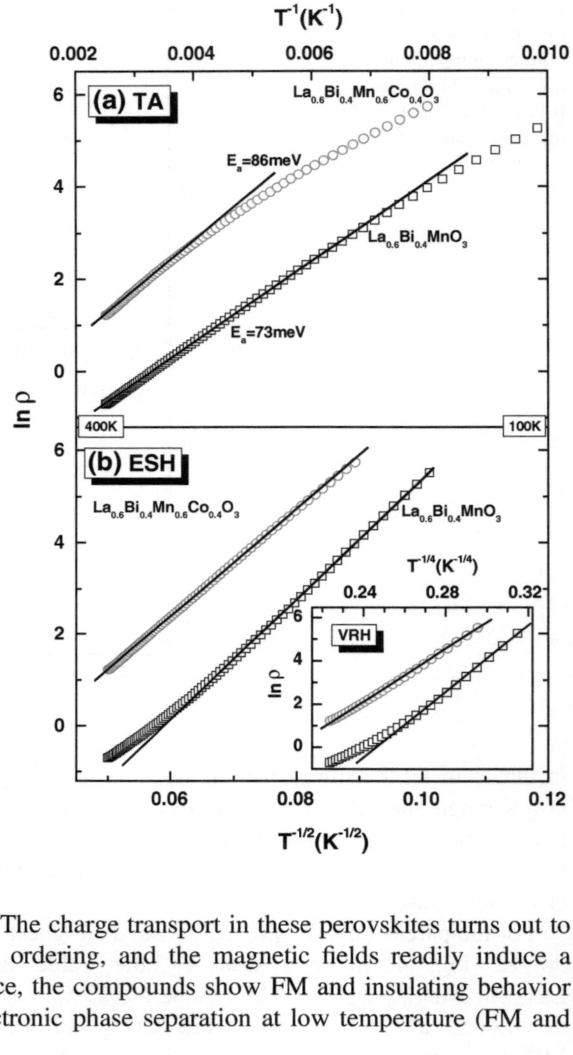

Fig. 4.22 Logarithm of resistivity versus $T^{-1/n}$ (where n = 1, 2 or 4) plots for $La_{0.6}Bi_{0.4}MnO_3$ and $La_{0.6}Bi_{0.4}Mn_{0.6}Co_{0.4}O_3$ (adapted from Ref. [22])

presence of an external field. The charge transport in these perovskites turns out to be very sensitive to the FM ordering, and the magnetic fields readily induce a negative MR below T_C. Hence, the compounds show FM and insulating behavior (even at high field) with electronic phase separation at low temperature (FM and AFM clusters).

Different models [56] have been proposed to describe the temperature-dependent charge transport behavior for magnetic perovskites at low temperature defined by ln $\rho \propto T^{-1/n}$, where n = 1, 2 or 4. Here, n = 1, corresponds to a simple thermal activation (TA) model, when n = 2, the hopping is referred to as *Efros-Shklovskii*-type hopping (ESH), controlled by Coulombic forces. When n = 4, there would be variable range hopping (VRH) and the hopping dynamics is controlled by collective excitation of the charge carriers. These models are compared to describe the zero field $\rho(T)$ behavior of the magnetic perovskites.

Figure 4.22a shows the TA model, ln $\rho \propto T^{-1}$, which describes the resistivity behavior above the FM transition T_C with the activation energy, E_a, of 73 and 86 meV for $La_{0.6}Bi_{0.4}MnO_3$ and $La_{0.6}Bi_{0.4}Mn_{0.6}Co_{0.4}O_3$, respectively. This suggests that with increasing the doping concentration at Mn-site the energy band gap increases gradually, similar to the Bi-centered magnetic perovskites [57]. It is noticed from Fig. 4.22b that, the ESH model, ln $\rho \propto T^{-1/2}$, suitably describes the resistivity behavior for Co-substituted phase in the whole measured temperature range (100 K ≤ T ≤ 400 K) among the three classes of models generally applied to the perovskite materials [58]. The ESH type conductivity is usually observed when the coulomb interaction starts to play a key role in carriers hopping. For Co-substituted phase, the resistivity evolution follows the VRH model only in the low temperature region ($T \leq 220$ K) as shown in the inset of Fig. 4.22b. This type of hopping conduction is typical to the phase separated systems (FM and AFM clusters) where the charge carrier moves by hopping between two localized electronic states. Similarly, the $La_{0.6}Bi_{0.4}Mn_{0.6}Fe_{0.4}O_3$ and $La_{0.6}Bi_{0.4}Mn_{0.6}Ni_{0.4}O_3$ perovskites are FMI below room temperature and consequently the MR effect is small in magnitude. The TA model, ln $\rho \propto T^{-1}$, describes well the resistivity behavior of the samples above the FM transition T_C with the activation energy, E_a, of 77 and 82 meV for $La_{0.5}Bi_{0.4}Mn_{0.6}Fe_{0.4}O_3$ and $La_{0.6}Bi_{0.4}Mn_{0.6}Ni_{0.4}O_3$, respectively, and are in good agreement with the $La_{0.6}Bi_{0.4}MnO_3$ and $La_{0.6}Bi_{0.4}Mn_{0.6}Co_{0.4}O_3$ compounds [22, 24]. This suggests that as the doping concentration at Mn-site is increased, the energy band gap gradually increases. The SPH model, ln $\rho\ \alpha\ T^{-1/2}$, describes the resistivity behavior for the $La_{0.6}Bi_{0.4}Mn_{0.6}Ni_{0.4}O_3$ compound in the temperature range of 100 K ≤ T ≤ 400 K [24].

4.4.2 Magnetodielectric Properties of Bulk $La_{0.6}Bi_{0.4}Mn_{0.6}Co_{0.4}O_3$

The isothermal magnetodielectric ($\Delta\varepsilon$) effects for $La_{0.6}Bi_{0.4}Mn_{0.6}Co_{0.4}O_3$ at different temperatures are discussed in this section. A positive magnetodielectric effect of 0.21–0.25 % in both the real and imaginary dielectric constant is observed (Fig. 4.23a) in the temperature range of 10–80 K, similar to $La_{0.5}Bi_{0.5}Mn_{0.67}Co_{0.33}O_3$ thin film [20]. Above 90 K, the $\Delta\varepsilon$ displays a gradual increase from 0.5 to 10 % up to 300 K. The difference in sign of variation of the ε' and ε'' dielectric constant above 90 K is observed, which signifies that the variation in capacitance and loss with magnetic field arises due to the presence of local regions (grain and grain boundaries) with varying resistivity [50].

Figure 4.23b shows the variation of ε' and ε'' with temperature, and the ε' value varies from 30 to 3000 in the temperature range of 10–300 K. At lower temperatures (≤80 K), both the ε' and ε'' values are independent of frequency as well as temperature. Around 140–200 K, a rapid increase of frequency-dependent dielectric constant behavior is observed. This type of sudden increase in the dielectric constant behavior is reported in the literature for other magnetic perovskites [42, 59]. The rapid increase in the dielectric constant values (two orders of magnitude)

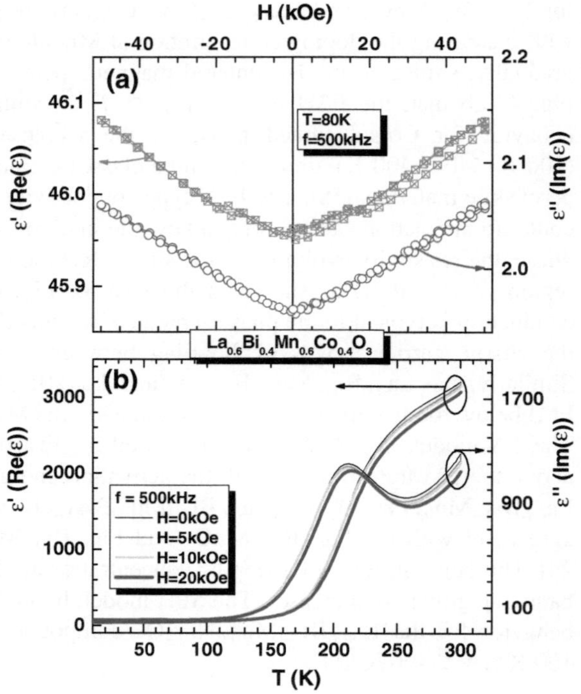

Fig. 4.23 **a** Magnetic field dependent isothermal real (ε') and imaginary (ε'') parts of dielectric constants at 80 K and **b** Temperature-dependent ε' and ε'' values in different magnetic fields for $La_{0.6}Bi_{0.4}Mn_{0.6}Co_{0.4}O_3$ at a frequency of 500 kHz (adapted from Ref. [22])

around 150 K arise from various factors, like intrinsic to the system and/or extrinsic effects. In the case of intrinsic effect, the steep rise of the dielectric constant over a short range of temperature could arise around the onset of ferroelectric ordering in which the dipoles are subjected to a double-well energy barrier. In that case, the dielectric constant obeys the Curie–Weiss behavior above the transition temperature, which is not the case in $La_{0.6}Bi_{0.4}Mn_{0.6}Co_{0.4}O_3$. The effect is due to the grain boundary and at temperatures above 250 K a combined effect of the grain boundary and the space charge effect arising due to the semiconducting behavior associated with the contact effects.

Figure 4.24 shows the frequency dispersion of ε' and ε'' at five different temperatures. The frequency-dependent ε' exhibits a large value at low frequency (<100 Hz) and a step like decrease with increase in frequency (>1 kHz) at 130 K. An associated *Debye* like relaxation peak in ε'' in the vicinity of the step-like frequency dispersion is observed in ε'. The observed relaxation peaks shift toward higher frequencies with increasing temperature similar to an activation behavior. The frequency-dependent behavior of the real and imaginary part of the dielectric constant is commonly observed in the case of materials exhibiting colossal dielectric behavior ($\varepsilon' > 1000$) due to extrinsic effects [42, 59].

Though the relaxation behavior looks like a *Debye*-type relaxation (a characteristic of the intrinsic dipoles), the temperature-dependent frequency shift suggests

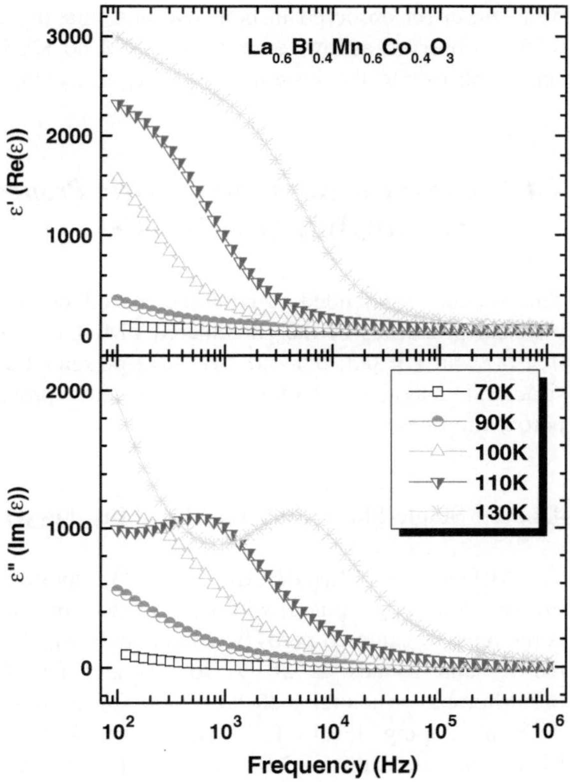

Fig. 4.24 Frequency dispersion behavior of the real (ε') and imaginary (ε'') part of the dielectric constants at different temperatures for $La_{0.6}Bi_{0.4}Mn_{0.6}Co_{0.4}O_3$ (adapted from Ref. [22])

that the relaxation is of *Maxwell–Wagner* type, which arises due to the presence of regions with different conductivities within the sample [42, 50, 59]. The large values in the low-frequency range of ε' imply that the grain boundary capacitance is larger than the bulk grain capacitance of the sample. Hence, the charge accumulated at the grain boundaries could give rise to a *Maxwell–Wagner*-type relaxation phenomenon [42, 59]. The gradual increase of the ε' value at low frequency (Fig. 4.23b) (for $T \geq 250$ K) is due to the space charge (due to DC conduction) and other contact-based effects.

The capacitance at different temperatures and frequencies reveal the overlap of the grain boundary relaxation of the sample. Hence, it is important to select the suitable temperature and the frequency to understand the magnetocapacitance effect reported for poor insulating samples. It is observed that the resistivity value increases rapidly (Fig. 4.17b) with decreasing temperature, hence, for $T < T_C$ carrier effects are not expected to play a relevant role in dielectric properties. Therefore, the capacitance measured at low temperature (T < 90 K) and high frequencies (f > 500 kHz) provides true intrinsic effects of the sample [50]. The $\Delta\varepsilon$ is 0.5–10 % for $La_{0.6}Bi_{0.4}Mn_{0.6}Co_{0.4}O_3$, and is dominated by the grain boundary conduction effect above 100 K for frequency range (100 Hz–1 MHz). Moreover, the positive weak $\Delta\varepsilon$ effect observed in both real and imaginary dielectric constants of 0.21–0.25 % observed at low temperatures ($T \leq 80$ K), is due to the spin lattice interaction intrinsic to the system akin to $La_{0.5}Bi_{0.5}Mn_{0.67}Co_{0.33}O_3$ thin film [20].

4.4.3 Magnetotransport-Dielectric Properties of Bulk $La_{1-x}Bi_xMn_{1-y}Fe_yO_3$ Series

This section is intended to correlate the effect of compositional changes on the physical properties of the Bi-centered FMI $La_{1-x}Bi_xMn_{1-y}Fe_yO_3$, where $x = 0.2$, 0.5, 0.7 and $y = 0.3$, 0.5–1.0. We have presented a comprehensive discussion on structural, magnetic, electrical, and dielectric properties of different Bi-centered perovskite phases.

4.4.3.1 Single-Phase Bulk $La_{0.5}Bi_{0.5}Mn_{0.5}Fe_{0.5}O_3$

The XRD pattern of $La_{0.5}Bi_{0.5}Mn_{0.5}Fe_{0.5}O_3$ reported by Jha et al. [19] indicates that the sample is single phase, without any traces of impurities (Fig. 4.25). In fact, this is the only Bi-centered perovskite manganite which shows the doping level in both the *A*- and *B*-sites as 50 % for Bi and Fe. The diffraction pattern shows orthorhombic structure, with the *Pnma* space group, similar to other phases with different doping levels [18–24]. Figure 4.26 shows the room temperature Mössbauer spectrum for $La_{0.5}Bi_{0.5}Mn_{0.5}Fe_{0.5}O_3$, which consists of a quadrupolar doublet indicating the PM behavior. The spectrum analysis reveals the presence of

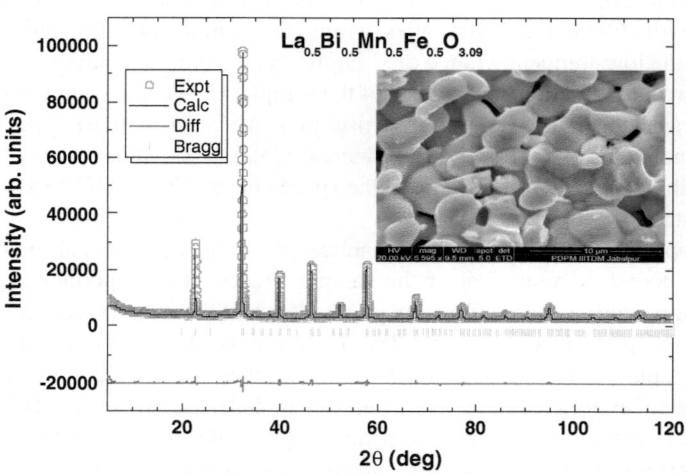

Fig. 4.25 Rietveld analysis of the XRD pattern at room temperature and *inset figure* shows the SEM images for $La_{0.5}Bi_{0.5}Mn_{0.5}Fe_{0.5}O_3$ (adapted from Ref. [19])

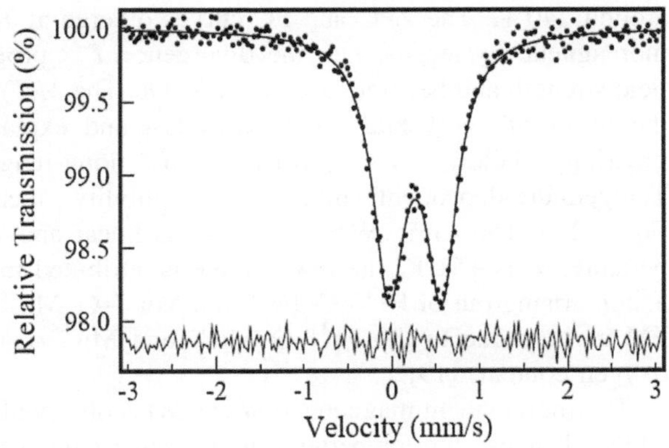

Fig. 4.26 Mossbauer spectra for $La_{0.5}Bi_{0.5}Mn_{0.5}Fe_{0.5}O_3$ at room temperature (adapted from Ref. [19])

one iron Mössbauer site with isomer shift of 0.38(1) mm/s and the quadrupole splitting value $\Delta E = 0.55(1)$ mm/s [19]. The values are expected for Fe^{3+} ion in high spin state and comparable to the values reported for perovskites $LaFeO_3$ (Ref. [60]) and $LaMn_{0.5}Fe_{0.5}O_3$ (Ref. [61]).

Figure 4.27a shows the ZFC and FC magnetization at three different applied fields. With decreasing temperature a PM to weak FM transition has been observed

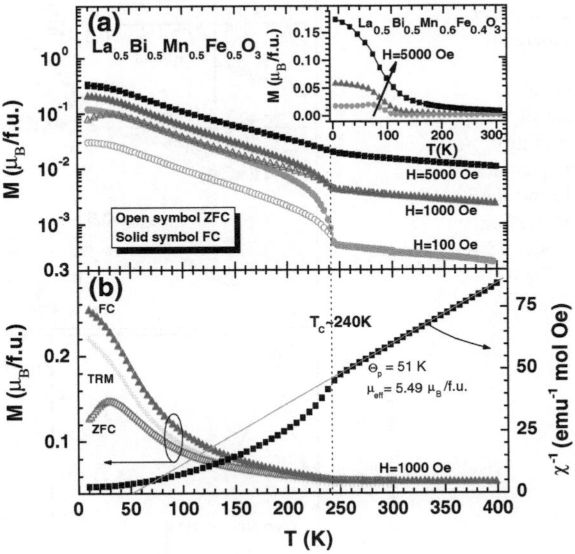

Fig. 4.27 Temperature-dependent ZFC (*open symbols*) and FC (*solid symbols*) magnetization, *M* (*T*), for different applied fields **a** H = 100, 1000 and 5000 Oe. **b** Thermoremanent magnetization (TRM) along with ZFC–FC magnetization (H = 1000 Oe) and the inverse magnetic susceptibility, χ^{-1}, versus temperature plot for $La_{0.5}Bi_{0.5}Mn_{0.5}Fe_{0.5}O_3$. *Inset figure* shows the *M*(*T*) for $La_{0.5}Bi_{0.5}Mn_{0.6}Fe_{0.4}O_3$ in three different applied fields (adapted from Ref. [19])

around 240 K. The ZFC and FC curves diverge at T_C ($T_{irr} \sim 240$ K). The thermomagnetic hysteresis, i.e., the divergence T_{irr}, progressively decrease with the field strength and becomes zero at 5000 Oe. The *M*(*T*) behavior partly corroborates the result of perovskite $LaMn_{0.5}Fe_{0.5}O_3$, and explained by the local magnetic ordering below T_C, instead of long-range FM ordering [61]. Temperature-dependent inverse susceptibility measurements are shown in Fig. 4.27b. The Curie–Weiss behavior is linear above 250 K and the Curie temperature, θ_p is +51 K; the lower value is attributed to the presence of AFM interaction arising out of Fe^{3+}–O–Fe^{3+} and Mn^{3+}–O–Mn^{3+} clusters coexisting with the FM clusters due to the Fe^{3+}–O–Mn^{3+} and Mn^{3+}–O–Mn^{4+} (related to the excess oxygen content) [19].

Thermoremanent magnetization (TRM) is observed at low temperature ($T < T_C$), which changes with temperature in a manner similar to the difference between the FC and ZFC magnetization. With increasing the field value, T_{irr} as well as the ZFC peak shifts toward lower temperature similar to that of a spin or cluster glass type material [32, 33], which is also supported by the unsaturated *M*(*H*) behavior (Fig. 4.28a).

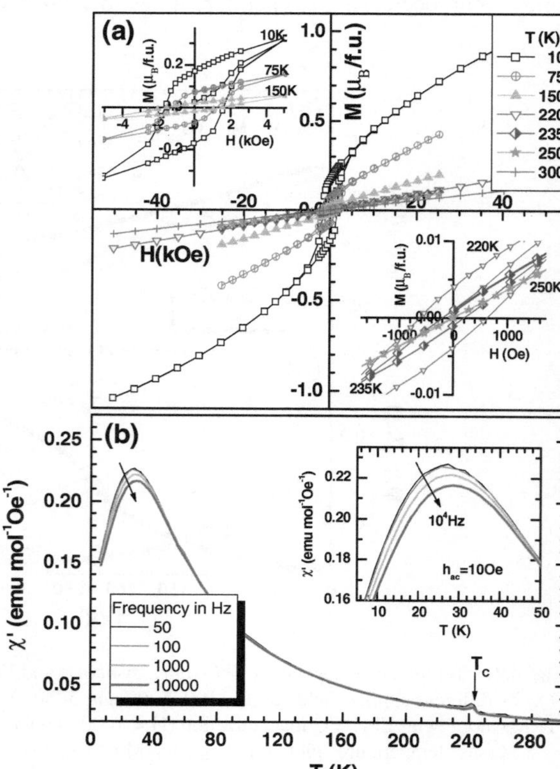

Fig. 4.28 **a** Magnetic field dependent isothermal magnetization, *M*(*H*), at seven different temperatures (*inset figures* show the enlarged version). **b** Temperature-dependent in-phase component of magnetic ac-susceptibility, χ', at four different frequencies (h_{ac} = 10 Oe) and *inset figure* shows the enlarged version around ZFC cusp (near 28 K) for $La_{0.5}Bi_{0.5}Mn_{0.5}Fe_{0.5}O_3$ (adapted from Ref. [19])

Figure 4.28b shows the temperature-dependent in-phase, χ', component of the AC susceptibility measured at different frequencies. The χ' data exhibits features similar to the ZFC magnetization data. There is a weak anomaly corresponding to FM T_C, which is frequency-independent and a frequency-dependent broad maximum corresponding to ZFC cusp (around 28 K). The latter shifts toward higher temperature with increasing frequency (inset of Fig. 4.28b), which is a characteristic feature of spin glasses [32, 33]. The FM and AFM components are competing with each other at low temperature due to the presence of Mn^{3+} and Fe^{3+} ions similar to $LaMn_{0.5}Fe_{0.5}O_3$ system [61]. The Fe^{3+}–O–Mn^{3+} superexchange (SE) interactions are indeed responsible for the FM component, characterized by a finite value of coercive field with hysteresis loops. Nevertheless, the μ_{eff} value ($\sim$5.49 μ_B/f.u.) is smaller than the theoretical spin-only value for Fe^{3+} and Mn^{3+} ions in high-spin states and the value of PM Curie temperature (θ_p $\sim$ 51 K) is lower than T_C, indicating the presence of strong AFM interactions due to Fe^{3+}–O–Fe^{3+} and Mn^{3+}–O–Mn^{3+} interactions. Consequently, the system is electronically phase separated into FM and AFM domains, giving rise to a glassy-FM state [30, 31].

The magnetic features of this system are considerably complex in nature. If the system is purely FM then it supposed to exhibit even sharp magnetic transition and saturation magnetization below magnetic ordering with increasing field strength. But that is not the reality. However, the thermomagnetic irreversibility, TRM effect and ZFC cusp signify glassy FM state with short-range FM ordering. A similar type of local short-range ordering below T_C has been reported in $LaMn_{0.5}Fe_{0.5}O_3$ phase [61]. In $La_{0.5}Bi_{0.5}Mn_{0.5}Fe_{0.5}O_3$, both the iron and manganese are in +3 oxidation state as mentioned earlier. The crystalline electric field lead to splitting of $3d^5$ state of Fe^{3+} to $t_{2g}^3e_g^2$ and is expected to provide a regular octahedron. Whereas Mn^{3+} is a *Jahn–Teller* active center ($3d^4$: $t_{2g}^3e_g^1$) and leads to distorted octahedron. The Mn^{3+} magnetic center in the distorted octahedra will interact anisotropically with its neighboring symmetric octahedral having Fe^{3+} moments. This 1:1 alternate regular and distorted octahedral centers, may give rise to the anisotropic interaction leading to helical arrangement of the spin [19].

The resistivity behavior for $La_{0.5}Bi_{0.5}Mn_{0.5}Fe_{0.5}O_3$, is similar to the previously discussed perovskites, gradually increases with decreasing temperature. The ln ρ versus T^{-1} plot is perfectly linear which signifies the TA model. The resistivity behavior in the 120–400 K temperature range leads to the activation energy, E_a of 80 meV. This is consistent with the value of Bi-centered perovskites [22, 24]. The Seebeck coefficient *S*(*T*) data signifies insulating type behavior in the 180–400 K range. The *S*(*T*) at room temperature is around +85 μV/K, and it increases gradually with decreasing temperature. The positive value of thermopower signifies the *p*-type polaronic conductivity or hole-like carriers in the materials. This is consistent with the behavior discussed earlier for Bi-centered perovskites (inset Fig. 4.4). Likewise, the MR values are negative and the highest value is about −5 % at 130 K, and near the FM ordering the value is ∼−1 %. Although the MR values are less than the expected value in the FM region, yet there is a definite correlation between the magnetic and electronic states at low temperature regions. With increasing temperature, the MR value decreases gradually and finally near the room

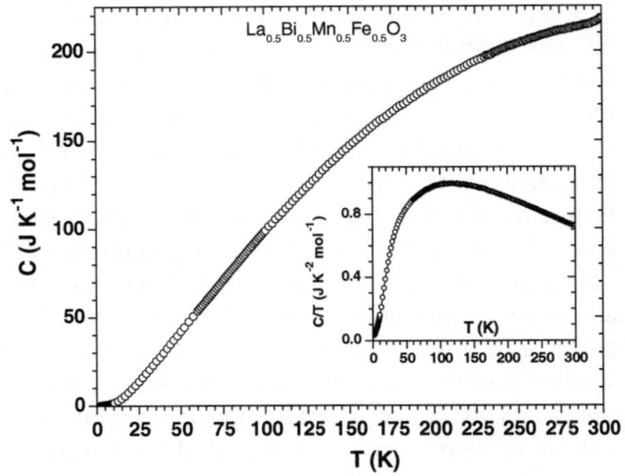

Fig. 4.29 The plots of heat capacity (*C*) as a function of temperature and *inset figure* shows *C*/*T* versus temperature for $La_{0.5}Bi_{0.5}Mn_{0.5}Fe_{0.5}O_3$ (adapted from Ref. [19])

temperature the value is ~ -0.2 %. In the case of a helical or noncollinear magnetic structure the spins are inherently involved in competition between themselves and hence their response to the external field is not expected to be very high [19].

The temperature dependence of heat capacity for $La_{0.5}Bi_{0.5}Mn_{0.5}Fe_{0.5}O_3$ is shown in Fig. 4.29. The characteristic λ-anomaly or a peak is absent in the specific heat *C*(*T*) plot (Fig. 4.29), thereby confirming the absence of long-range AFM or FM interaction [19]. There is a gradual decrease in the value with decreasing temperature and the fall is nonlinear in certain regions. This feature is more clearly visible in the *C*/*T* versus temperature plot (inset of Fig. 4.29). The *C*(*T*) shows two distinct slope changes in the behavior around 10 and 175 K. However, the trend of *C*(*T*) is similar to silicon below room temperature with a higher value of *C*(*T*) [19].

Figure 4.30 shows the dielectric behavior of $La_{0.5}Bi_{0.5}Mn_{0.5}Fe_{0.5}O_3$ in the frequency range of 10 kHz–1 MHz. The ε' is varying from 20 to 600 and the ε'' varies from 0.01 to 6000. The rise in the dielectric constant to a colossal magnitude around 120 K associated with a local peak in ε'' (indicated by a downward arrow around 160 K in Fig. 4.30) is due to extrinsic effects such as grain boundaries (*Maxwell–Wagner* type), twin boundaries (if any present in the system), and/or other conductive effects [42, 50]. Above 230 K, the electrode material interface and space charge effects collectively dominate the capacitance measurements and display a monotonous rise in the magnitudes of both real and imaginary permittivity [21]. A weak positive magnetodielectric effect of around 0.15–0.25 % for the temperature range 10–80 K is reported, and attributed to an intrinsic effect of the compound [5]. Whereas a large magnetodielectric effect of around 0.3–10 % is reported in the range of 100–300 K, with a maximum around 220–250 K, due to other extrinsic effects, such as grain boundary conductance [50].

The inset of Fig. 4.30 shows a magnified view of ε' in a narrow temperature range of 50–72 K. A weak anomaly in the ε', observed at $\sim$62 and $\sim$65 K in the ε'', becomes predominant in the presence of magnetic field, shown in the inset of Fig. 4.30. This originates from the onset of polar behavior. A theoretical study on

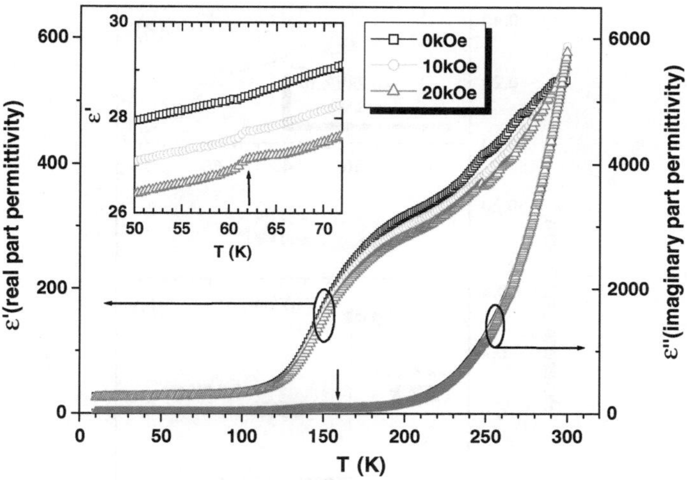

Fig. 4.30 Temperature-dependent real (ε') and imaginary (ε'') part of permittivity in different magnetic fields for $La_{0.5}Bi_{0.5}Mn_{0.5}Fe_{0.5}O_3$. *Inset figure* shows the magnified view near the weak dielectric anomaly (around 62 K) (adapted from Ref. [19])

magnetic perovskite reveals that the plausible polar ground states can couple with an octahedral lattice, consisting of different magnetic cations [12]. The presence of $6s^2$ lone pair electrons in the Bi^{3+} cation builds an additional structural distortion into the lattice, explaining the anomaly observed in the dielectric studies.

Figure 4.31a shows the temperature-dependent electric polarization in an applied electric field of ±3.2 kV/cm, which exhibits a polar behavior with a remnant polarization 0.3 μC/cm^2 at 10 K. A polar-to-nonpolar kind of transition is observed around 67 K [19]. The polarization values have been reported to increase with the applied electric field and become saturated at 0.3 μC/cm^2 at an electric field above 3.2 kV/cm. The reversal of polarization is shown in Fig. 4.31b and is achieved successively down to 20 K (step 3). The polarization behavior and its reversal nature were noticed both in the heating and cooling cycles of the sample. The dielectric anomaly associated with a drop in polarization reveals the polar behavior of the compound below 67 K. The observed polarization is also reversible with applied field and effectively satisfies the ferroelectric criteria [62]. The coexistence of both FM and FE behavior proves the multiferroic nature of $La_{0.5}Bi_{0.5}Mn_{0.5}Fe_{0.5}O_3$ phase [19].

4.4.3.2 Single-Phase Bulk $La_{0.5}Bi_{0.5}Mn_{0.7}Fe_{0.3}O_3$, $La_{0.5}Bi_{0.5}Mn_{0.3}Fe_{0.7}O_3$ and $La_{0.3}Bi_{0.7}Mn_{0.3}Fe_{0.7}O_3$

Figure 4.32 shows the ZFC and FC magnetization, *M*(*T*), for $La_{0.5}Bi_{0.5}Mn_{0.7}Fe_{0.3}O_3$, $La_{0.5}Bi_{0.5}Mn_{0.3}Fe_{0.7}O_3$ and $La_{0.3}Bi_{0.7}Mn_{0.3}Fe_{0.7}O_3$. Unlike

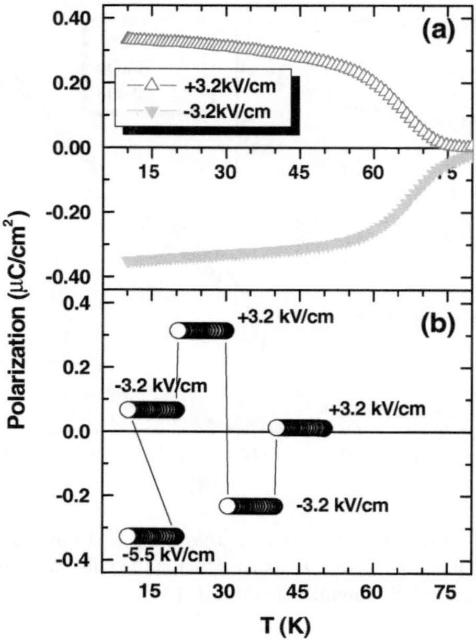

Fig. 4.31 **a** Temperature-dependent electric polarization and **b** Electric field dependent polarization reversal for $La_{0.5}Bi_{0.5}Mn_{0.5}Fe_{0.5}O_3$ (adapted from Ref. [19])

the multiferroic phase $La_{0.5}Bi_{0.5}Mn_{0.5}Fe_{0.5}O_3$ (Ref. [19]) having weak FM behavior, the Mn-rich phase $La_{0.5}Bi_{0.5}Mn_{0.7}Fe_{0.3}O_3$ exhibits $T_C \sim 110$ K. On the contrary, heavily Fe-substituted phases $La_{0.5}Bi_{0.5}Mn_{0.3}Fe_{0.7}O_3$ and even with the modification of *A*-site composition, i.e., $La_{0.3}Bi_{0.7}Mn_{0.3}Fe_{0.7}O_3$ leads to significant change in FM behavior and becomes non-FM type (Fig. 4.32b, c). For $La_{0.5}Bi_{0.5}Mn_{0.7}Fe_{0.3}O_3$, the ZFC and FC exhibit large divergence below ~ 50 K. This indicates that the magnetic interactions in this system are different from the long-range ferromagnetism [25]. For Fe-rich systems, a gradual increase in the M (T) is reported below 100 K, which is attributed to the weak FM or canted AFM ordering similar to $BiFeO_3$ [4, 63]. The $M(H)$ loops are shown in the insets of Fig. 4.32. A prominent hysteresis loop with a remanent magnetization (M_r) of ~ 0.2 μ_B/f.u. and a coercive field (H_C) of ~ 1720 Oe is observed for $La_{0.5}Bi_{0.5}Mn_{0.7}Fe_{0.3}O_3$. In contrast, the coercive fields for $La_{0.5}Bi_{0.5}Mn_{0.3}Fe_{0.7}O_3$ and $La_{0.3}Bi_{0.7}Mn_{0.3}Fe_{0.7}O_3$ are 1980 and 2450 Oe, respectively, at 10 K (inset of Fig. 4.32b, c), which are relatively higher than that of the Mn-rich phase. None of these compositions exhibit saturation even at 5 T. The highest value of magnetic moment, ~ 1.2 μ_B/f.u. is reported for $La_{0.5}Bi_{0.5}Mn_{0.7}Fe_{0.3}O_3$, whereas for $La_{0.5}Bi_{0.5}Mn_{0.3}Fe_{0.7}O_3$ and $La_{0.3}Bi_{0.7}Mn_{0.3}Fe_{0.7}O_3$ the corresponding values are much lower. Such lower moment is counterintuitive, since substitution of Fe for Mn changes d^4 ($S = 2$) cation to d^5 ($S = 2.5$). Actually, the smaller values of moment for $La_{0.5}Bi_{0.5}Mn_{0.3}Fe_{0.7}O_3$ and $La_{0.3}Bi_{0.7}Mn_{0.3}Fe_{0.7}O_3$, has been attributed to the

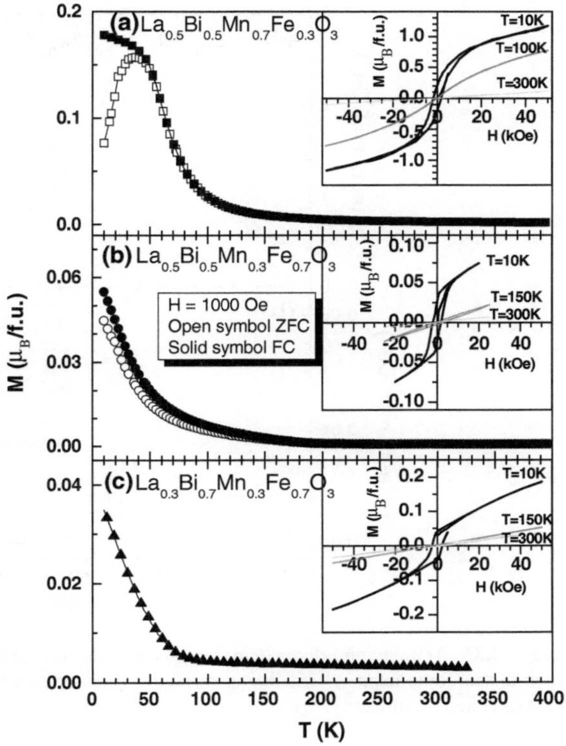

Fig. 4.32 Temperature-dependent ZFC (*open symbol*) and FC (*solid symbol*) magnetization, *M*(*T*), and *M* (*H*) loops at different temperatures for **a** $La_{0.5}Bi_{0.5}Mn_{0.7}Fe_{0.3}O_3$, **b** $La_{0.5}Bi_{0.5}Mn_{0.3}Fe_{0.7}O_3$, and **c** $La_{0.3}Bi_{0.7}Mn_{0.3}Fe_{0.7}O_3$ (adapted from Ref. [25])

change in the mode of magnetic interaction in the Fe-rich phases, i.e., the FM interactions between Fe^{3+}–O–Mn^{3+} ions are weakened by the strong Fe^{3+}–O–Fe^{3+} AFM interaction [25]. Though the Mn-rich phase shows a clear FM T_C, the *M* (*H*) behavior is not as is expected for FM system, rather it is akin to glassy ferromagnets [30, 31]. The AC magnetization (Fig. 4.33) at four different frequencies for $La_{0.5}Bi_{0.5}Mn_{0.7}Fe_{0.3}O_3$ shows the glassy behavior. At low temperatures there is a subtle balance between the FM and AFM interactions or in other words the system is electronically phase separated into FM and AFM clusters, giving rise to glassy FM state [30, 31].

Figure 4.34 shows the $\rho(T)$ for $La_{0.5}Bi_{0.5}Mn_{0.7}Fe_{0.3}O_3$, $La_{0.5}Bi_{0.5}Mn_{0.3}Fe_{0.7}O_3$ and $La_{0.3}Bi_{0.7}Mn_{0.3}Fe_{0.7}O_3$ perovskites. The resistivity value increases gradually with decreasing temperature. It is noticed that the samples are insulating throughout the measured temperature range (100 K ≤ T ≤ 400 K). Thus, similar to the other Bi-centered perovskites [18–25], none of these samples show any insulator–metal transition in the 100–400 K range. The ferroelectric behavior is established by the temperature-dependent dielectric measurements for the temperature range of 25–350 K.

The dielectric constant and dielectric loss for $La_{0.5}Bi_{0.5}Mn_{0.7}Fe_{0.3}O_3$, $La_{0.5}Bi_{0.5}Mn_{0.3}Fe_{0.7}O_3$ and $La_{0.3}Bi_{0.7}Mn_{0.3}Fe_{0.7}O_3$ are presented in Figs. 4.35 and 4.36, respectively. The dielectric value increases gradually with increasing

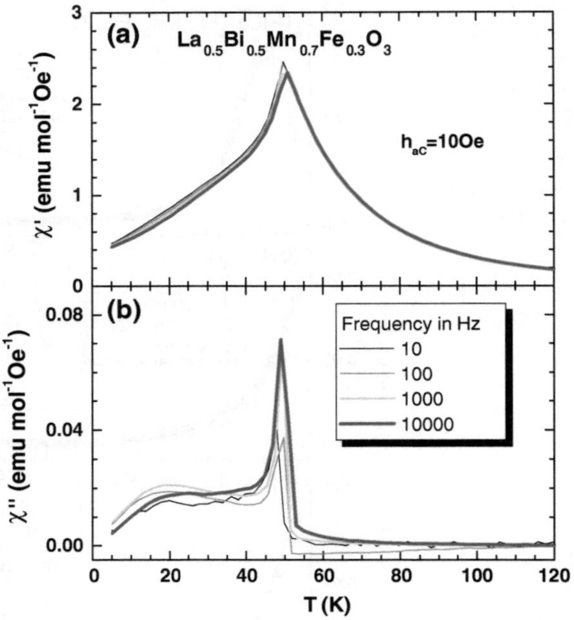

Fig. 4.33 Temperature-dependent magnetic ac-susceptibility for $La_{0.5}Bi_{0.5}Mn_{0.7}Fe_{0.3}O_3$ **a** in-phase component, χ' and **b** out-of-phase χ'' components at different frequencies (h_{ac} = 10 Oe) (adapted from Ref. [25])

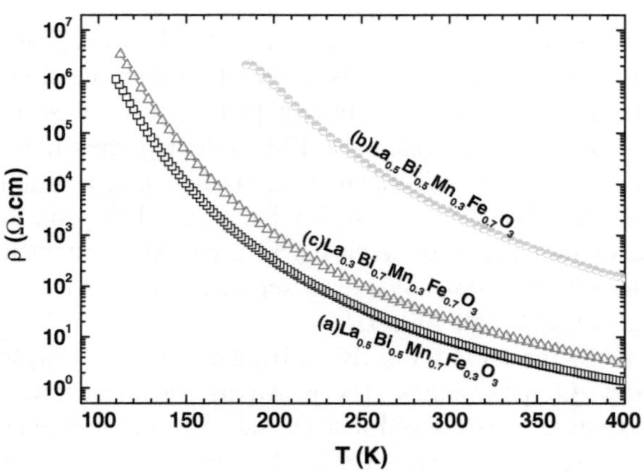

Fig. 4.34 Temperature-dependent electrical resistivity, $\rho(T)$, for **a** $La_{0.5}Bi_{0.5}Mn_{0.7}Fe_{0.3}O_3$ (*square symbol*) **b** $La_{0.5}Bi_{0.5}Mn_{0.3}Fe_{0.7}O_3$ (*circular symbol*) and **c** $La_{0.3}Bi_{0.7}Mn_{0.3}Fe_{0.7}O_3$ (*triangular symbol*) (adapted from Ref. [25])

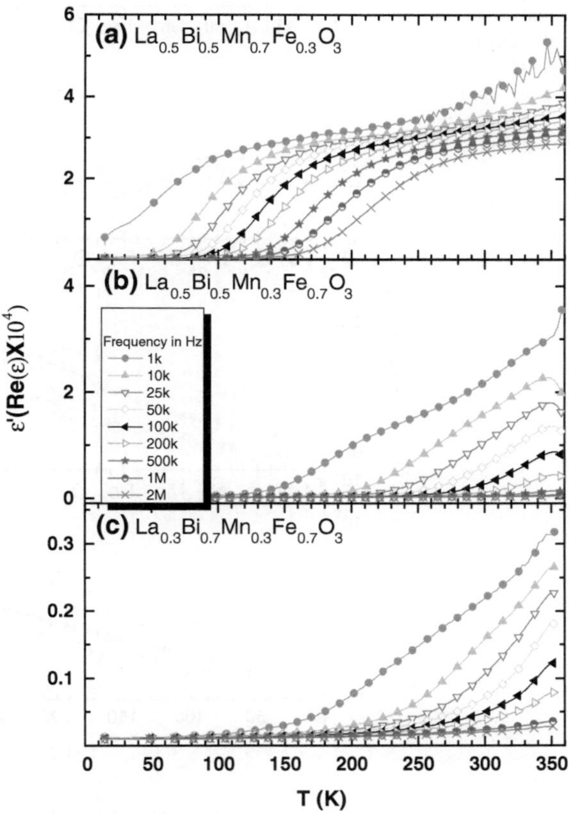

Fig. 4.35 Temperature-dependent dielectric constants for **a** $La_{0.5}Bi_{0.5}Mn_{0.7}Fe_{0.3}O_3$, **b** $La_{0.5}Bi_{0.5}Mn_{0.3}Fe_{0.7}O_3$ and **c** $La_{0.3}Bi_{0.7}Mn_{0.3}Fe_{0.7}O_3$ at different frequencies (adapted from Ref. [25])

temperature for the three compounds. In the case of $La_{0.5}Bi_{0.5}Mn_{0.7}Fe_{0.3}O_3$, the real part of the dielectric data exhibit plateaus with a change in slope (Fig. 4.35a). A gradual increase of frequency-dependent behavior is observed in the high temperature regions. The dielectric response of this system demonstrates relaxor-like behavior, i.e., the magnitude of the dielectric constant decreasing with increasing frequency. Giant dielectric constant (up to 32,000 at 100 kHz) is reported near room temperature. With decreasing temperature, the dielectric constant value rapidly decreases to a lower value (<2000 at 100 kHz).

A broad dielectric loss peak (Fig. 4.36a) is evident corresponding to the rapid change in region of the dielectric constant. The loss peak shifts to higher temperature as frequency increases (Fig. 4.36a), indicating a thermally activated relaxation. The dielectric loss increases significantly with increasing temperature due to the contribution of DC conductivity in the system [25].

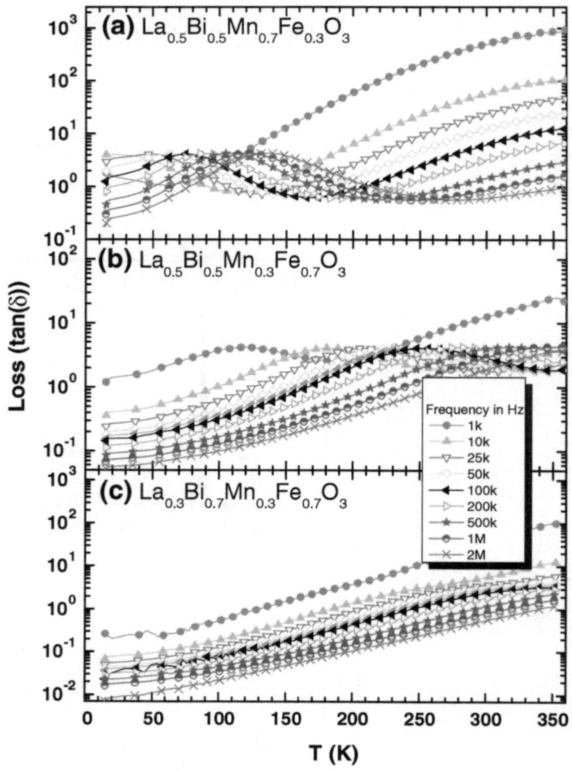

Fig. 4.36 Temperature dependence dielectric loss for **a** $La_{0.5}Bi_{0.5}Mn_{0.7}Fe_{0.3}O_3$, **b** $La_{0.5}Bi_{0.5}Mn_{0.3}Fe_{0.7}O_3$ and **c** $La_{0.3}Bi_{0.7}Mn_{0.3}Fe_{0.7}O_3$ at different frequencies (adapted from Ref. [25])

However, the real part of dielectric data for $La_{0.5}Bi_{0.5}Mn_{0.3}Fe_{0.7}O_3$ (Fig. 4.35b) and $La_{0.3}Bi_{0.7}Mn_{0.3}Fe_{0.7}O_3$ (Fig. 4.35c) do not show any plateaus and the value increases rapidly with increasing temperature (T > 130 K). For $La_{0.5}Bi_{0.5}Mn_{0.3}Fe_{0.7}O_3$, a dielectric maximum at around 345 K is observed (Fig. 4.35b), which shifts toward the high temperature (356 K) with increasing frequency (200 kHz), but this is absent for $La_{0.3}Bi_{0.7}Mn_{0.3}Fe_{0.7}O_3$ (Fig. 4.35c). The systems reveal that the series is described by dipolar-type relaxation along with variable type hopping conduction of the charge carriers, similar to other compounds [22, 23]. It is also well known that in the semiconducting-type materials, localized charge carriers hopping between spatially fluctuating lattice potentials not only produce conductivity but also give rise to dipolar effect. For the Fe-rich samples, around 150–350 K a rapid increase of frequency-dependent dielectric constant behavior is observed. Similar kind of sudden increase in the dielectric constant behavior is reported in the literature for Fe-doped systems [23]. For $La_{0.5}Bi_{0.5}Mn_{0.3}Fe_{0.7}O_3$ and $La_{0.3}Bi_{0.7}Mn_{0.3}Fe_{0.7}O_3$ phases the rapid increase in the dielectric constant values above 150 K (Fig. 4.35b, c) could arise from various factors, like intrinsic to the

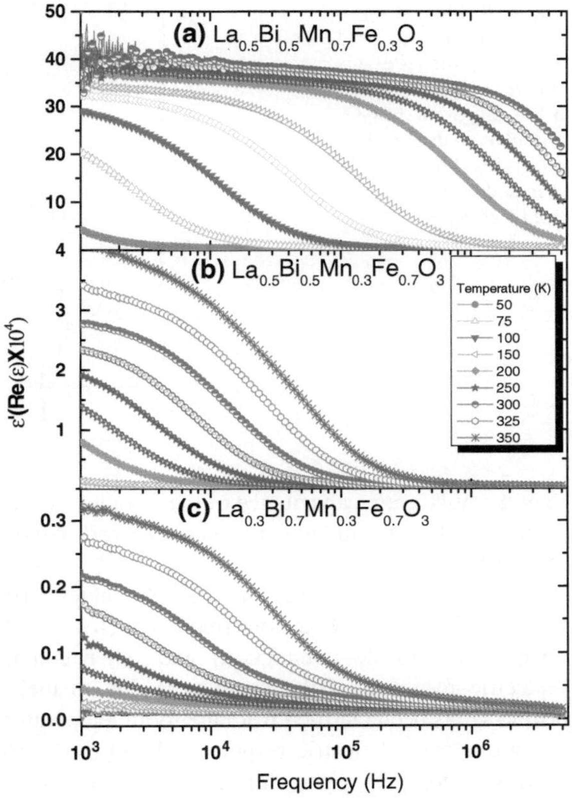

Fig. 4.37 Frequency dependence of dielectric constants at different temperatures for **a** $La_{0.5}Bi_{0.5}Mn_{0.7}Fe_{0.3}O_3$, **b** $La_{0.5}Bi_{0.5}Mn_{0.3}Fe_{0.7}O_3$ and **c** $La_{0.3}Bi_{0.7}Mn_{0.3}Fe_{0.7}O_3$ (adapted from Ref. [25])

system and/or extrinsic effects. The effect is due to the grain boundary and at temperatures above 250 K a combined effect of the grain boundary and the space charge effect as discussed for semiconducting samples [21, 25, 42, 50]. Nevertheless, for $La_{0.5}Bi_{0.5}Mn_{0.3}Fe_{0.7}O_3$ the transition peaks are prominent and remain well defined at high frequencies, i.e., the ferroelectricity in this system is intrinsic, similar to multiferroics $La_{0.8}Bi_{0.2}Fe_{1-x}Mn_xO_3$ and discussed later.

Figure 4.37 shows the frequency dispersion for the $La_{0.5}Bi_{0.5}Mn_{0.7}Fe_{0.3}O_3$, $La_{0.5}Bi_{0.5}Mn_{0.3}Fe_{0.7}O_3$ and $La_{0.3}Bi_{0.7}Mn_{0.3}Fe_{0.7}O_3$ compounds at different temperatures. The frequency-dependent ε' exhibits a large value at low frequency (<1 kHz) and a step like decrease with increase in frequency (>1 kHz) at 75 K for $La_{0.5}Bi_{0.5}Mn_{0.7}Fe_{0.3}O_3$ (Fig. 4.37a) and around 250 K for $La_{0.5}Bi_{0.5}Mn_{0.3}Fe_{0.7}O_3$ (Fig. 4.37b) and $La_{0.3}Bi_{0.7}Mn_{0.3}Fe_{0.7}O_3$ (Fig. 4.37c). The relaxation shifts toward higher frequencies with increasing temperature suggesting the activation behavior. The grain boundaries present and/or different magnetic clusters act as a region of different conductivity as observed in the case of magnetic properties. The large values in the low frequency ε' imply that the grain boundary capacitance is larger

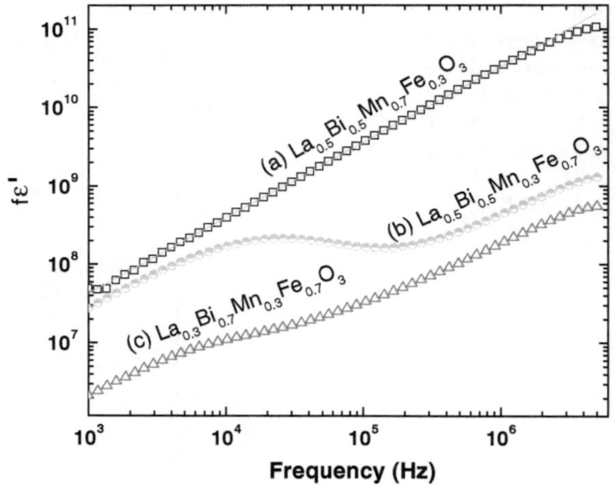

Fig. 4.38 Plots of log *f* versus log (*fε*′) at 300 K for **a** $La_{0.5}Bi_{0.5}Mn_{0.7}Fe_{0.3}O_3$ (*square symbol*), **b** $La_{0.5}Bi_{0.5}Mn_{0.3}Fe_{0.7}O_3$ (*circular symbol*) and **c** $La_{0.3}Bi_{0.7}Mn_{0.3}Fe_{0.7}O_3$ (*triangular symbol*) (adapted from Ref. [25])

than the bulk grain capacitance of the sample. Hence, the charge accumulated at the grain boundaries could give rise to a *Maxwell–Wagner* type relaxation phenomenon [42, 50].

The gradual increase of the ε' values for $La_{0.5}Bi_{0.5}Mn_{0.3}Fe_{0.7}O_3$ and $La_{0.3}Bi_{0.7}Mn_{0.3}Fe_{0.7}O_3$ at low frequency (for $T > 250$ K) arises due to the space charge (due to dc conduction) and other contact-based effects. Whether the dielectric response of the samples is due to their FE nature or by some other artifacts, has been clarified from the room temperature frequency-dependent data by using universal dielectric response (UDR) model [64]. Figure 4.38 shows the log f versus log $f\varepsilon'$ plots for all the three samples. The fitting curve for $La_{0.5}Bi_{0.5}Mn_{0.7}Fe_{0.3}O_3$ follows a linear behavior throughout the frequency range, whereas for $La_{0.5}Bi_{0.5}Mn_{0.3}Fe_{0.7}O_3$ and $La_{0.3}Bi_{0.7}Mn_{0.3}Fe_{0.7}O_3$ the linear regions are limited to certain frequency regime. Therefore, according to UDR model, the dielectric response of $La_{0.5}Bi_{0.5}Mn_{0.7}Fe_{0.3}O_3$ sample is intrinsic and for other two Fe-rich systems the response are combined effects [25].

4.4.3.3 Single-Phase Bulk $La_{0.8}Bi_{0.2}Mn_{1-y}Fe_yO_3$

Figure 4.39 shows the temperature-dependent magnetization for $La_{0.8}Bi_{0.2}Mn_{0.4}Fe_{0.6}O_3$, $La_{0.8}Bi_{0.2}Mn_{0.3}Fe_{0.7}O_3$ and $La_{0.8}Bi_{0.2}Mn_{0.1}Fe_{0.9}O_3$. The last phase shows complex magnetic phenomena, whereas the first two compounds show weak FM behavior, due to the presence of Mn in +2, +3, and +4 valence states. The magnetization value is higher for Mn-rich phase compared to Fe. This is due to mismatch of two magnetic ions Fe/Mn, which distorts the Fe/Mn octahedra causing frustration [23].

The electrical conductivity increases with increasing Mn content due to hopping of electrons from lower to higher valence states. A competition between AFM/FM

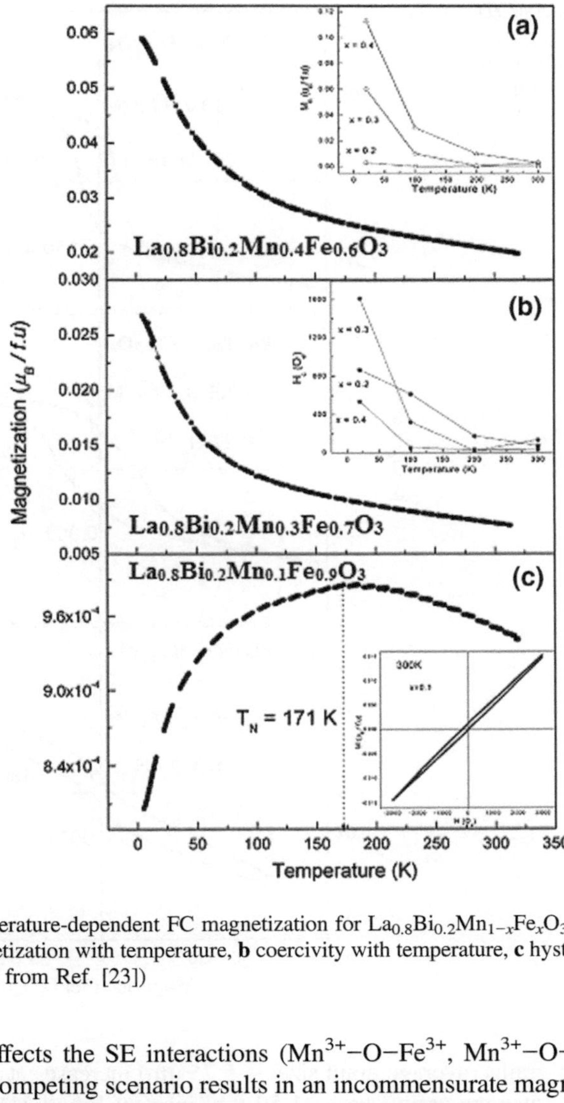

Fig. 4.39 Temperature-dependent FC magnetization for $La_{0.8}Bi_{0.2}Mn_{1-x}Fe_xO_3$ perovskites. *Inset* **a** remnant magnetization with temperature, **b** coercivity with temperature, **c** hysteresis behavior for $x = 0.9$ (adapted from Ref. [23])

lattices also affects the SE interactions (Mn^{3+}–O–Fe^{3+}, Mn^{3+}–O–Mn^{3+}, Fe^{3+}–O–Fe^{3+}). This competing scenario results in an incommensurate magnetic ordering in the system, which leads to an uncompensated (canted) magnetic interface that gives rise to weak FM behavior.

Figure 4.40 shows the *M*(*H*) loops for $La_{0.8}Bi_{0.2}Mn_{0.4}Fe_{0.6}O_3$, $La_{0.8}Bi_{0.2}Mn_{0.3}Fe_{0.7}O_3$ and $La_{0.8}Bi_{0.2}Mn_{0.2}Fe_{0.8}O_3$ at 20 K. The unsaturated behavior of hysteresis loop indicates canted nature of spins. Figure 4.41 shows that the overall grain size increases for Mn-rich phases. The system is composed of the distribution of two different types of grains. As the Mn content is increased, the

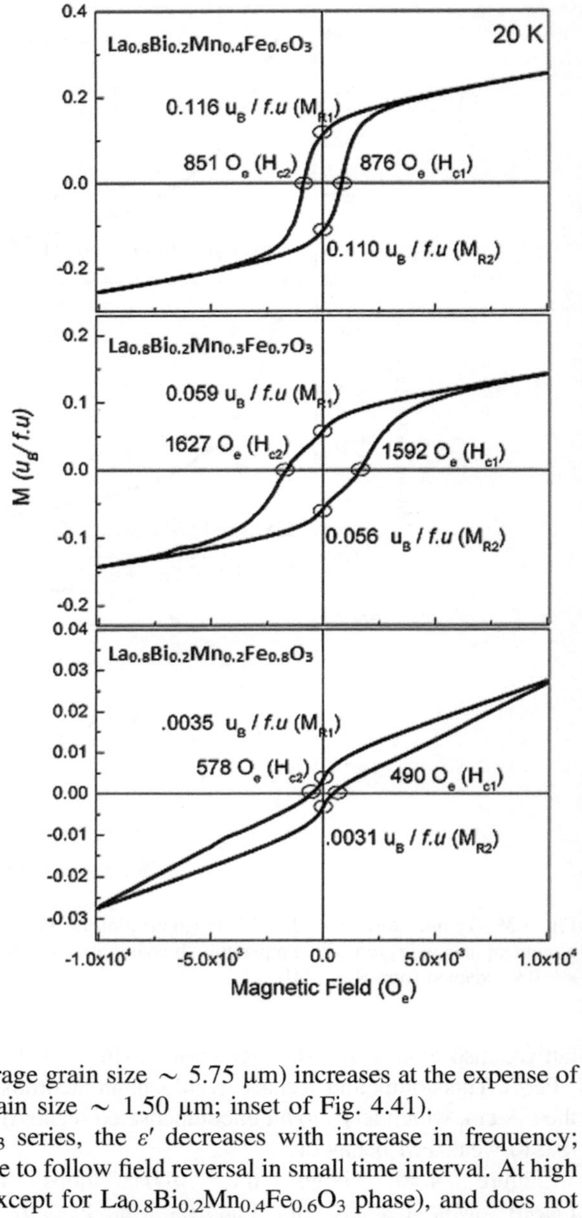

Fig. 4.40 Isothermal $M(H)$ hysteresis loops of $La_{0.8}Bi_{0.2}Mn_{1-y}Fe_yO_3$ at 20 K (adapted from Ref. [23])

number of larger grains (average grain size $\sim$ 5.75 μm) increases at the expense of smaller particles (average grain size $\sim$ 1.50 μm; inset of Fig. 4.41).

For $La_{0.8}Bi_{0.2}Mn_{1-y}Fe_yO_3$ series, the ε' decreases with increase in frequency; because the dipoles are unable to follow field reversal in small time interval. At high frequencies, ε' is constant (except for $La_{0.8}Bi_{0.2}Mn_{0.4}Fe_{0.6}O_3$ phase), and does not vary with frequency (Fig. 4.42), because the electrodes and grain boundaries are not influential when frequency is greater than 500 Hz. Dielectric property enhanced for lower value of Mn-ion at *B*-site and induces ferroelectricity due to Bi-ion at *A*-site. Magnetoelectric (ME) coupling reported for a wide range of temperature (180–280 K). The ME coupling for the $La_{0.8}Bi_{0.2}Mn_{0.3}Fe_{0.7}O_3$ composition is about

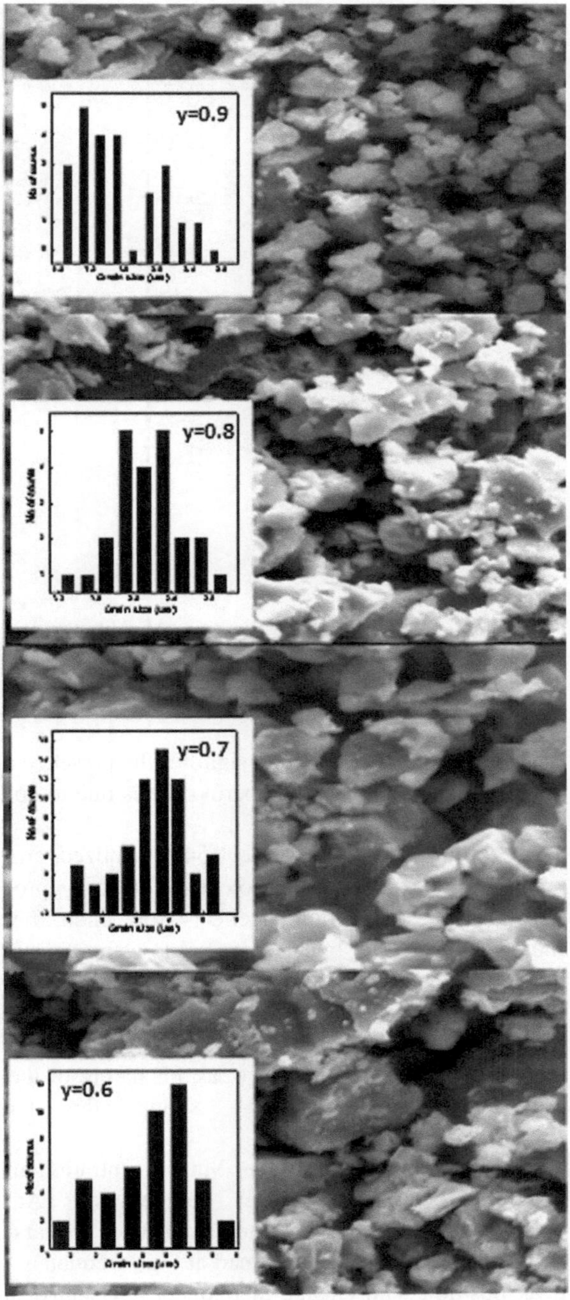

Fig. 4.41 FESEM micrographs of $La_{0.8}Bi_{0.2}Mn_{1-y}Fe_yO_3$ series taken on the scale of 10 μm. *Inset* show the respective histograms of grain sizes (adapted from Ref. [23])

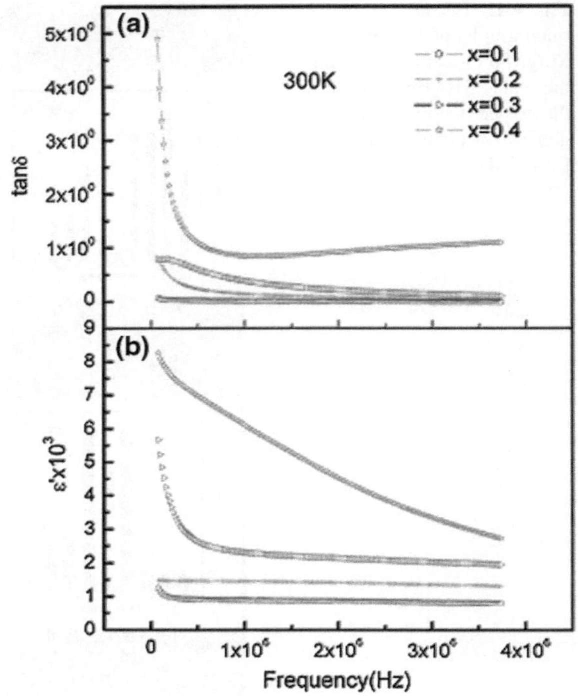

Fig. 4.42 **a** Variation of dielectric loss (tan δ) versus frequency at 300 K, and **b** dielectric constant (ε') of $La_{0.8}Bi_{0.2}Mn_{1-x}Fe_xO_3$ as a function of frequency (adapted from Ref. [23])

18 % at around 200 K, which signifies the presence of better ME coupling [23]. The magnetocapacitance for this perovskite is due to coupling between the electric and magnetic dipoles.

According to UDR Model [64], localized charge carriers hopping between spatially fluctuating lattice potentials not only produce the conductivity but also may give rise to the dipolar effects, for which, the plot between log f and log $f\varepsilon'$ must be linear. The linear nature at low frequency ($f < 500$ Hz) establishes UDR phenomenon as the guiding phenomenon for dielectric response. However, at higher frequency, the nonlinear nature (Fig. 4.43) rules out the contribution of electrode, grain boundary or *Maxwell–Wagner* effect to the dielectric response. The dielectric effect is due to the weak FE nature of the perovskites in high-frequency region [23]. Figure 4.44 shows the temperature-dependent ε' for $La_{0.8}Bi_{0.2}Mn_{1-y}Fe_yO_3$ series.

The ε' increases with higher Mn concentration and the main dielectric transition peak shifts toward lower temperature. At higher Mn concentration ($La_{0.8}Bi_{0.2}Mn_{0.4}Fe_{0.6}O_3$), a new transition appears along with primary dielectric peak. The peaks are well defined at high frequency (1 MHz), with slight change in peak position with change in frequency. Usually, at higher frequencies, transition peak does not remain well defined if the ferroelectricity in the system is due to

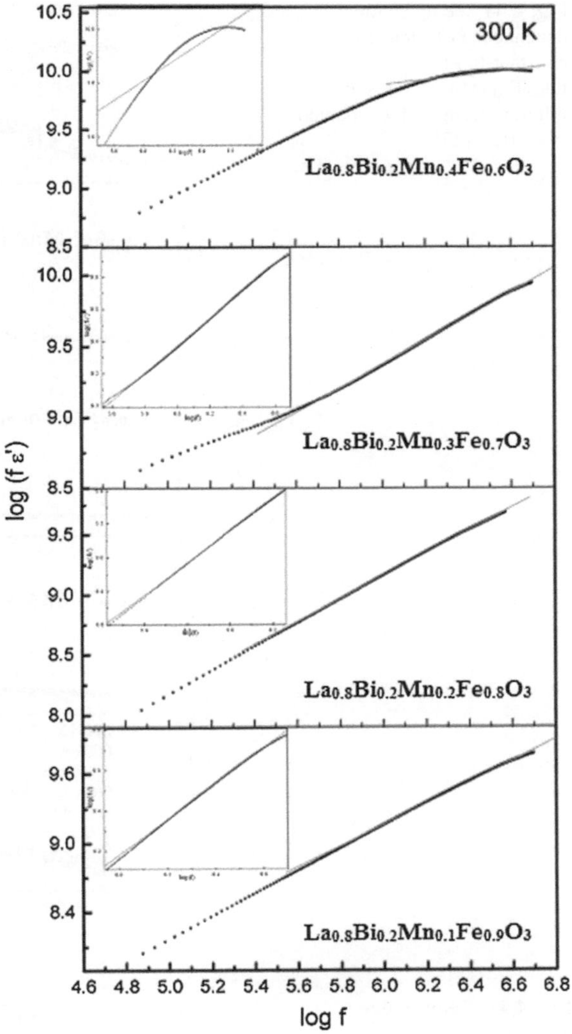

Fig. 4.43 Plots of log f versus log $f\varepsilon'$ for $La_{0.8}Bi_{0.2}Mn_{1-y}Fe_yO_3$ at 300 K in the frequency range of 75 kHz–4 MHz. *Inset* enlarged view of the linear fit (adapted from Ref. [23])

electrode, *Maxwell–Wagner* effect or grain boundary [42, 50]. The temperature-dependent dielectric loss (tan δ) is similar to the dielectric behavior. The tan δ is independent of frequency at low temperature and shifts toward higher temperature with increasing frequency due to the presence of relaxor-type behavior. Further, the tan δ is higher for Mn-rich phase at any temperature and frequency, which is due to increase in DC conductivity with higher Mn concentration.

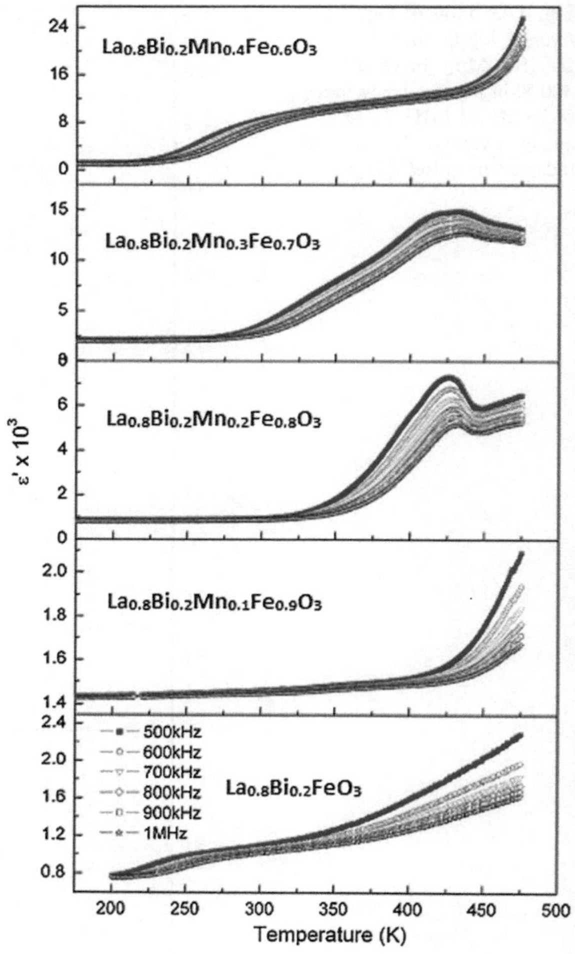

Fig. 4.44 Temperature dependence of dielectric constant (ε') for $La_{0.8}Bi_{0.2}Mn_{1-y}Fe_yO_3$ at different frequencies (adapted from Ref. [23])

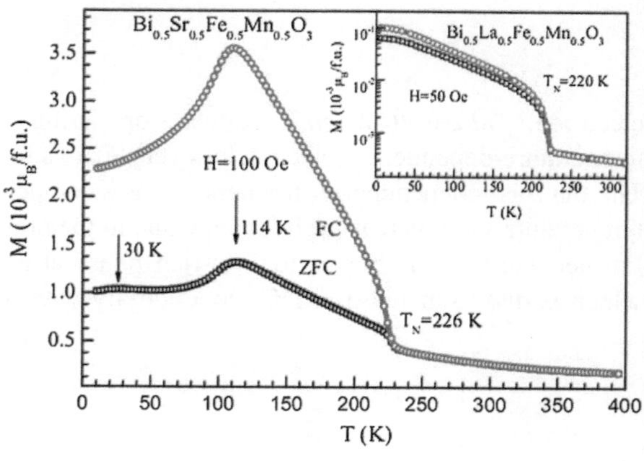

Fig. 4.45 Temperature dependent ZFC and FC Magnetization for $(La/Sr)_{0.5}Bi_{0.5}Mn_{0.5}Fe_{0.5}O_3$ (adapted from Ref. [26])

4.4.4 Magnetotransport Properties of Bulk (La/Sr/Bi)(MnFe/Cr)O_3 Perovskites

The magnetic properties of (La/Sr/Bi)(MnFe/Cr)O_3 perovskites are distinctly different from each other with weak FM/AFM ordering below room temperature [26]. Figure 4.45 shows the ZFC and FC magnetization, *M*(*T*), for $(La/Sr)_{1-x}Bi_xMn_{0.5}Fe_{0.5}O_3$ [26]. It may be mentioned here, except $BiFeO_3$, most of the other Bi based compounds are stabilized in the perovskite structure under high pressure and high temperature. The compound, $BiFe_{0.5}Mn_{0.5}O_3$, synthesized at high pressure and temperature, has an orthorhombic structure and exhibits a temperature-induced magnetization reversal below $T_N = 240$ K [10]. However, it is possible to stabilize La/Sr compound at ambient pressure by partial substitution at the *A*- or *B*-sites. For example, La or Sr substitutions at the Bi-site of the high-pressure $BiFe_{0.5}Mn_{0.5}O_3$ phase stabilize it in the perovskite structure at ambient pressure. In $La_{0.5}Bi_{0.5}Fe_{0.5}Mn_{0.5}O_3$, Fe^{3+} and Mn^{3+} orders antiferromagnetically at 240 K and this compound is a multiferroic and exhibits a spin glass state at low temperature as discussed previously. It is worthwhile to discuss the physical properties of two close systems with 50 % of La- and Sr-substituted at the Bi-site of $BiFe_{0.5}Mn_{0.5}O_3$, having d^5–d^4 (La doped) and d^5–d^3 (Sr doped) spin configuration, respectively. Interestingly, they differ in their magnetic properties. $La_{0.5}Bi_{0.5}Fe_{0.5}Mn_{0.5}O_3$ shows an AFM ordering with spin canting below $T_N = 220$ K, whereas $Sr_{0.5}Bi_{0.5}Fe_{0.5}Mn_{0.5}O_3$ exhibits a complex magnetic behavior. Below T_N (226 K), $Sr_{0.5}Bi_{0.5}Fe_{0.5}Mn_{0.5}O_3$ exhibits a cluster glass state and at further low temperature 30 K, a spin-glass state is observed [26]. Similarly, the perovskite $La_{0.5}Bi_{0.5}Fe_{0.5}Cr_{0.5}O_3$ shows the disordering of iron and chromium in the B-sites, similar to other discussed perovskites. More importantly, this perovskite is found to be an uncompensated weak ferromagnet, with a very peculiar zero magnetization behavior, generally observed for ordered magnetic cations in the B sites. It exhibits a magnetic transition at high temperature ($T_C \sim 450$ K), while the zero magnetization occurs between 100 and 160 K [26].

4.5 Bismuth-Centered Ordered Magnetic Perovskites $La_{2-x}Bi_xMn(Co/Ni)O_6$

In this section, we have discussed two series of Bi-centered ferromagnetic perovskites $La_{2-x}Bi_xMn(Co/Ni)O_6$, which are derived from ordered perovskite $La_2Mn(Co/Ni)O_6$. These perovskites $La_{2-x}Bi_xMn(Co/Ni)O_6$ could be synthesized at ambient pressure using sol–gel method [65–68]. Ricciardo et al. [65] have reported composition up to $x = 1.0$ with some impurities and the crystal structures are similar to the ordered perovskite $La_2Mn(Co/Ni)O_6$ [9, 26–28, 38, 39]. While Nautiyal et al. [66] have reported the presence of significant impurity phases for $La_{2-x}Bi_xMnNiO_6$ perovskites with $x \geq 0.4$. However, the higher Bi content composition can be

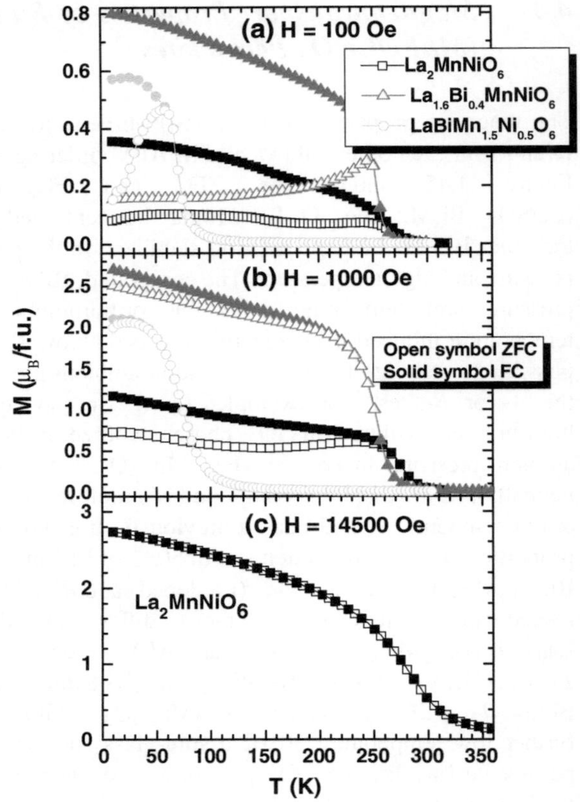

Fig. 4.46 Temperature-dependent ZFC (*open symbol*) and FC (*solid symbol*) magnetization, *M*(*T*), for La_2MnNiO_6 (*square*), $La_{1.6}Bi_{0.4}MnNiO_6$ (*triangle*) and $LaBiMn_{1.5}Ni_{0.5}O_6$ (*circle*), in an applied field of **a** H = 100 Oe; **b** H = 1000 Oe and **c** H = 14500 Oe (for La_2MnNiO_6) (adapted from Ref. [66])

obtained with single phase at ambient pressure by changing the Mn–Ni ratio, e.g., $LaBiMn_{1.5}Ni_{0.5}O_6$, but only in disordered perovskite phase [66]. On the other hand, Bai et al. [67] have reported single phase $La_{2-x}Bi_xMnCoO_6$, for $x = 0$, 0.1, 0.2 and 0.3 compositions with ordered structure.

4.5.1 Magnetic Properties

Figure 4.46a shows the PM to FM transition at 275 K for the ordered perovskite La_2MnNiO_6. The curve does not saturate below transition temperature, indicating that the magnetism is not truly long-range order. The Bi doping at *A*-site results into significant drop in T_C as reported for $La_{1.6}Bi_{0.4}MnNiO_6$ and $LaBiMn_{1.5}Ni_{0.5}O_6$ phases. The FM T_C values are 255 and 75 K, respectively, as shown in Fig. 4.46a, b. A large divergence between the ZFC and FC below the transition temperature is observed, similar to ordered La_2MnNiO_6 [99, 26–28]. This is due to the phase separation of FM domains distributed in an AFM matrix [66]. At higher applied

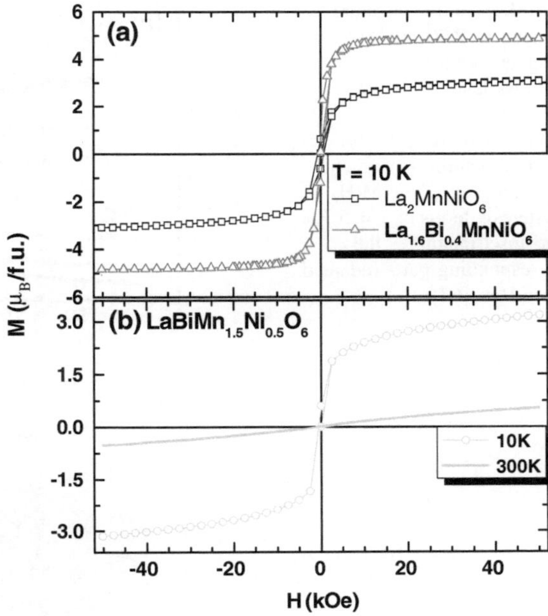

Fig. 4.47 Field dependent isothermal magnetic hysteresis, *M*(*H*), curves at different temperatures for **a** La_2MnNiO_6 (*square*) and $La_{1.6}Bi_{0.4}MnNiO_6$ (*triangle*) and **b** $LaBiMn_{1.5}Ni_{0.5}O_6$ (*circle*) (adapted from Ref. [66])

field, the ZFC and FC curves merged down to low temperatures. The *M*(*H*) loops (Fig. 4.47) signify soft FM nature. The magnetic moment for La_2MnNiO_6 is smaller than $La_{1.6}Bi_{0.4}MnNiO_6$ phase, although the magnetic interactions should be similar in nature, i.e., FM interactions between Mn^{4+}–O–Ni^{2+} [9, 26–28]. Ricciardo et al. [65] have proposed the lower values of moment due to varying degrees of site disorder, which introduce AFM nearest neighbor interactions. For $LaBiMn_{1.5}Ni_{0.5}O_6$ phase the smaller value of moment and T_C(≅75 K), compared to the other two phases is explained by its lower ratios of Mn^{4+} and Ni^{2+} ions. This is due to charge balance as $La^{3+}Bi^{3+}Mn^{3+}(Mn^{4+}Ni^{2+})_{0.5}O_6$ which induces a smaller Mn^{4+} content. Consequently, FM (due to Mn^{4+} and Ni^{2+} ions) and AFM (due to Mn^{3+} and Mn^{3+}) interactions would contribute significantly, and a subtle balance between FM and AFM interactions suppresses the FM T_C for $LaBiMn_{1.5}Ni_{0.5}O_6$ phase [66].

The magnetization *M*(*T*) and *M*(*H*) loops for $La_{2-x}Bi_xMnCoO_6$ series are shown in Fig. 4.48. The ZFC magnetization for La_2MnCoO_6 and $La_{1.9}Bi_{0.1}MnCoO_6$ phases is negative, whereas it is positive for $La_{1.7}Bi_{0.3}MnCoO_6$ phase. This has been attributed to the *B*-site cation induced formation of the anti-phase boundary [67].

The T_C decreases from 230 to 205 K as the Bi concentration is increased from 0 to 0.3. It seems intuitively adverse that a lower magnetic transition temperature is reported in a more ordered system [67]. This is unusual phenomena attributed to the fact that T_C and θ_p are more sensitive to the orbital overlapping geometry than to the degree of ordering in the matrix that is governed by the Co–Mn ordering [65, 67]. The decrease in the T_C is mainly recognized to the reduced Mn(Co)–O–Mn(Co)

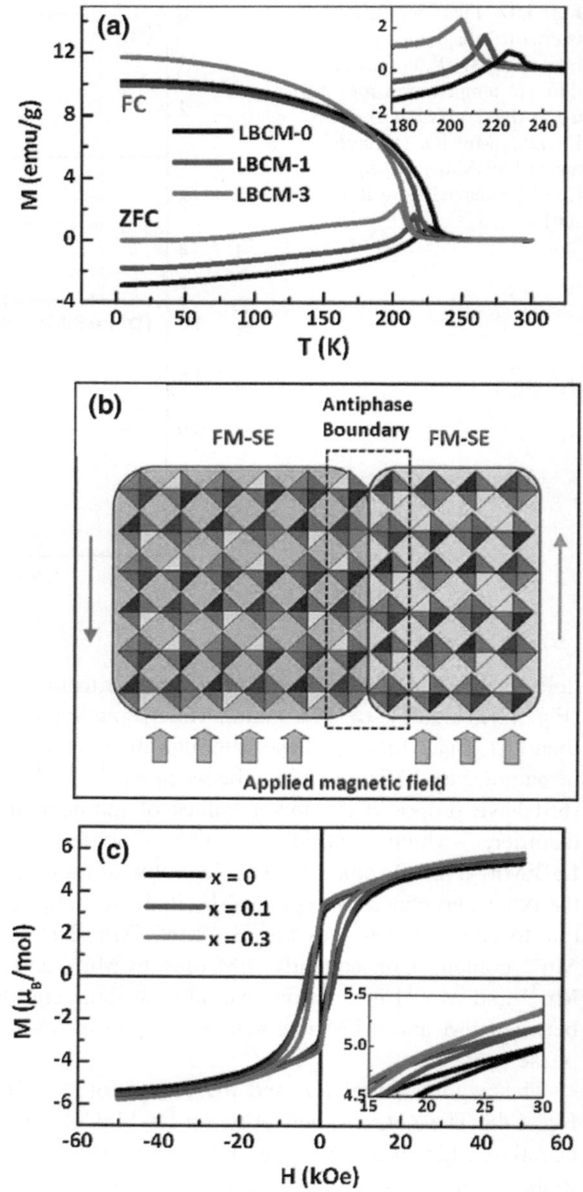

Fig. 4.48 **a** Temperature dependent ZFC and FC magnetization for perovskitess $La_{2-x}Bi_xMnCoO_6$. **b** Schematic diagram for microstructure of magnetic domains and **c** The M(H) hysteresis loops at $T = 5$ K and inset magnifies the corresponding parts (adapted from Ref. [67])

bond angle from 160.2(4)° to 157.0(6)° and the elongated Mn(Co)−O bond length from 1.973(7) to 1.986(9) Å, which suppress effectively the Co^{2+}−Mn^{4+} FM SE interactions via the O^{2-} by lowering the degree of orbital overlapping [67]. Bai et al. [67] proposed a special PM state comprising super-PM clusters and domains that is favorable near T_C.

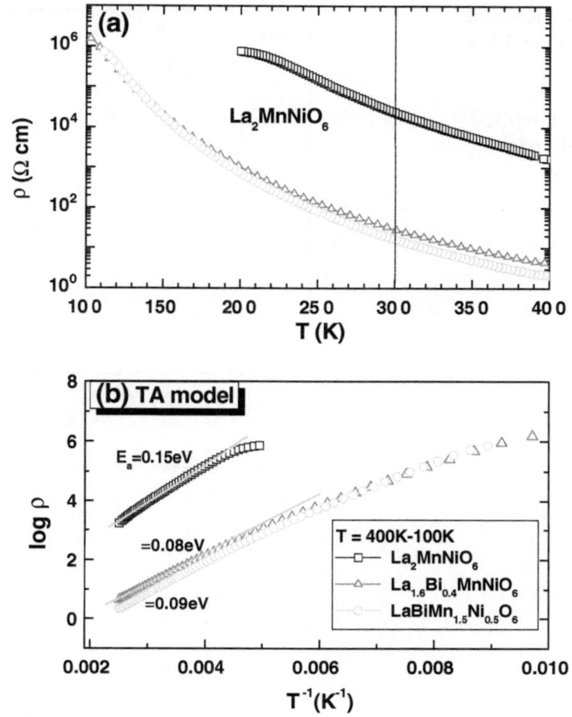

Fig. 4.49 **a** Temperature-dependent electrical resistivity, $\rho(T)$, and **b** Logarithm of resistivity versus inverse of temperature plots for La_2MnNiO_6 (*square*) $La_{1.6}Bi_{0.4}MnNiO_6$ (*triangle*) and $LaBiMn_{1.5}Ni_{0.5}O_6$ (*circle*) (adapted from Ref. [66])

4.5.2 Electrical Properties

With decreasing temperature, the resistivity increases for all samples and the values are very high at low temperature (Fig. 4.49a), signifying the insulating behavior [66]. In the 100–400 K temperature range, the $\rho(T)$ data confirms the insulating phase although the materials are FM below room temperature. The TA model (Fig. 4.49b) describes the zero field $\rho(T)$ behavior for the perovskites above the FM T_C with the activation energy, E_a, of 0.15, 0.08, and 0.09 eV for La_2MnNiO_6, $La_{1.6}Bi_{0.4}MnNiO_6$ and $LaBiMn_{1.5}Ni_{0.5}O_6$, respectively [66]. This suggests that with increasing the Bi substitutions at La-site, the energy band gap decreases. The DC resistivity is smaller for the Bi-substituted phases. Such an increase in the dc conductivity indeed denotes an increase in the grain interior conductivity of the Bi-doped phases [66].

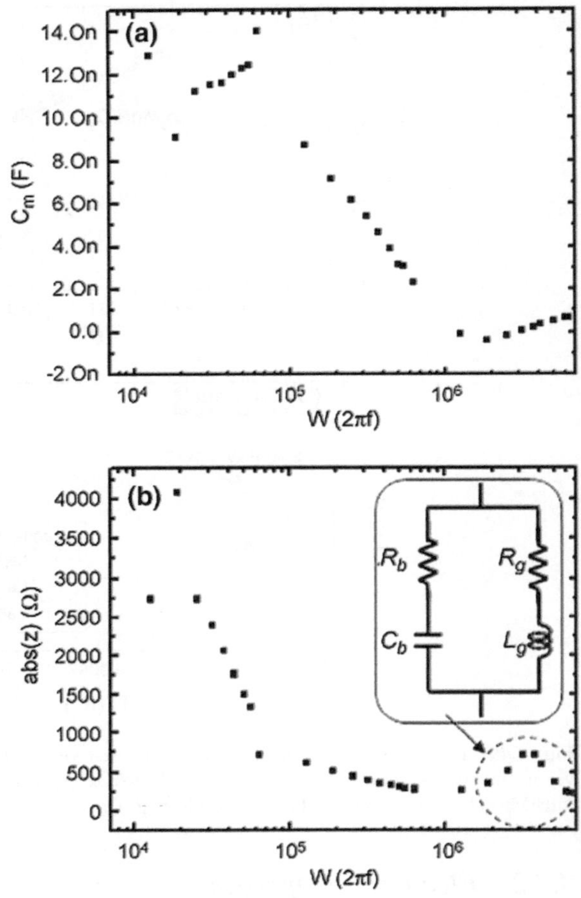

Fig. 4.50 Typical measured. **a** Capacitance (Cm) and **b** Z impedance of $La_{1.5}Bi_{0.5}MnCoO_6$ from 1 kHz to 1 MHz at 300 K (adapted from Ref. [65])

Figure 4.50 shows the capacitance and impedance with respect to frequency for $La_{1.5}Bi_{0.5}MnCoO_6$. As the frequency decreases, the capacitance significantly increases. The high value of capacitance at low frequencies implies the presence of ionic conductivity. At high frequency, the conductivity which persists at low temperatures is associated with electronic conductivity. The impedance curve in high-frequency region exhibits a resonance as in a parallel RLC resonant circuit [65]. By considering the resonant feature of the impedance at high frequencies, a simple analytic circuit model in the circuit point of view is established as shown in the inset of Fig. 4.50b.

Dielectric constant of the ionic conductor component can be assumed to be constant at high frequencies (>300 kHz) while it is frequency-dependent at low frequencies. In this model, R_b and C_b correspond to the resistance and capacitance

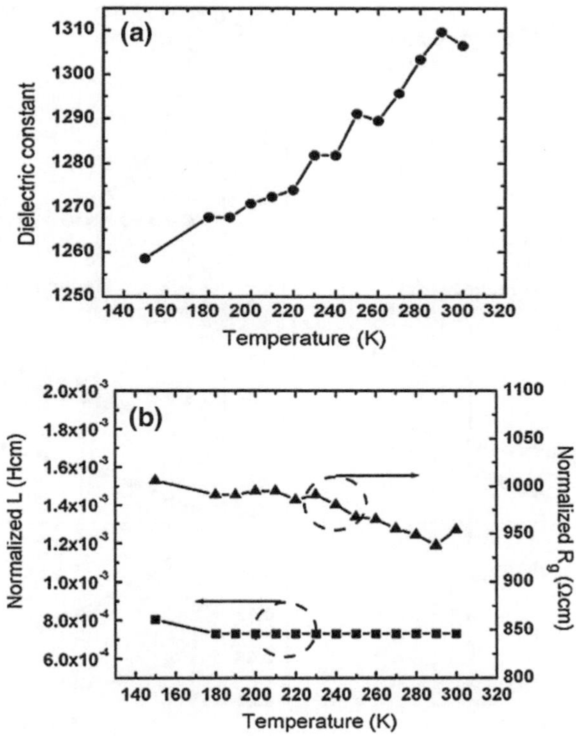

Fig. 4.51 Temperature-dependence of **a** dielectric constant and **b** normalized L_g and R_g calculated from the extracted C_b, L_g, and R_g of $La_{1.5}Bi_{0.5}MnCoO_6$ (adapted from Ref. [65])

of the bulk, and R_g and L_g are related to parasitic component along the grain boundaries [65]. Figure 4.51 shows the temperature dependence of dielectric constant, and normalized L_g and R_g of $La_{1.5}Bi_{0.5}MnCoO_6$, in high-frequency range of 300 kHz to 1 MHz. With increasing temperature dielectric constant increases, while R_g decreases and L_g remains constant. The large values of the dielectric constant reported to be come from polarization effects associated with conductivity, either in the grain boundary or in the bulk. However, there is no indication of an electrical transition in the region of the magnetic transition. Thus, there is no clear evidence of coupling between the magnetic and electrical properties in this Bi-centered ordered perovskite [65].

Figure 4.52 shows the frequency dispersion plot at room temperature and the temperature-dependent ε' and tan δ plots for $La_{2-x}Bi_xMnCoO_6$ series [68]. For high Bi-concentration the dielectric relaxation is negligible and the ε' and tan δ become weakly frequency—dependentat high frequencies. For La_2MnCoO_6 and

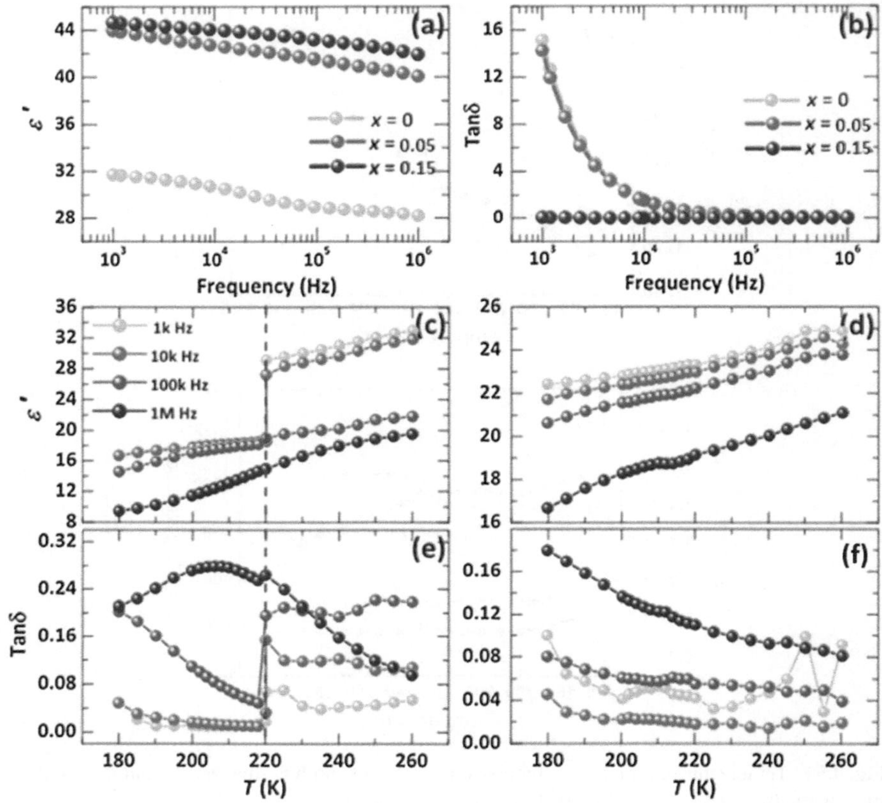

Fig. 4.52 Frequency dependence of **a** dielectric constant (ε') and **b** dielectric loss (tan δ) for the $La_{2-x}Bi_xMnCoO_6$ samples with $x = 0$, 0.05, and 0.15. The measurements were conducted at 300 K. The temperature dependence of the ε' for the $La_{2-x}Bi_xMnCoO_6$ with **c** $x = 0$ and **d** $x = 0.15$ under the frequencies of 1, 10, 100 kHz, and 1 MHz. Corresponding tan δ are given in the plots (**e**) and (**f**) (adapted from Ref. [68])

$La_{1.7}Bi_{0.3}MnCoO_6$ a step-like transition is observed in both the ε' and tan δ at around 220 K (Fig. 4.52c–f). The anomaly corresponds to a strong coupling between the magnetic and dielectric behavior as reported by Bai et al. [68].

References

1. P. K. Baltzer, H. W. Lahmann, M. Robbins, Phys. Rev. Lett. **15**, 493 (1965); G. A. Prinz, J. Mag. Mag. Mater. **200**, 57 (1999)
2. N.A. Hill, J. Phys. Chem. B **104**, 6694 (2000)

3. N. A. Hill, A. Filipetti, J. Magn. Magn. Mater. **242–245**, 976 (2002); A. M. Santos, S. Parashar, A. R. Raju, Y. S. Zhao, A. K. Cheetham, C. N. R. Rao, Solid State Commun. **122,** 49 (2002); A. M. Santos, A. K. Cheetham, T. Atou, Y. Syono, Y. Yamaguchi, K. Ohoyama, H. Chiba, C. N. R. Rao, Phys. Rev. B **66**, 064425 (2002)
4. J. Wang, J.B. Neaton, H. Zheng, V. Nagarajan, S.B. Ogale, B. Liu, D. Viehland, V. Vaithyanathan, D.G. Schlom, U.V. Waghmare, N.A. Spaldin, K.M. Rabe, M. Wuttig, R. Ramesh, Science **299**, 1719 (2003)
5. T. Kimura, S. Kawamoto, Y. Yamada, M. Azuma, M. Takano, Y. Tokura, Phys. Rev. B **67**, 180401(R) (2003)
6. D.V. Efremov, J. van den Brink, D.I. Khomskii, Nature Mater. **3**, 853 (2004)
7. M. Fiebig, J. Phys. D **38**, R123 (2005)
8. W. Prellier, M.P. Singh, P. Murugavel, J. Phys.: Condens. Matter **17**, R803 (2005)
9. N.S. Rogado, J. Li, A.W. Sleight, M.A. Subramanian, Adv. Mat. **17**, 2225 (2005)
10. G. T. Rado, Phys. Rev. Lett. **6**, 609 (1961); T. Kimura, T. Goto, H. Shintani, K. Ishizaka, T. Arima, Y. Tokura, Nature (London) **426**, 55 (2003); T. Goto et al., Phys. Rev. Lett. **92**, 257201 (2004); N. Hur, S. Park, P. A. Sharma, S. Guha, S. W. Cheong, Phys. Rev. Lett. **93**, 107207 (2004); C. R. Serrao et al., Phys. Rev. B, **72**, 220101(R) (2005); N. Ikeda et al., Nature, **436**, 1136 (2005); I. A. Sergienko, E. Dagotto, Phys. Rev. B, **73**, 094434 (2006); I. A. Sergienko, C. Sen, E. Dagotto, Phys. Rev. Lett. **97**, 227204 (2006); O. Heyer et al., J. Phys.: Condens. Matter **18**, L471 (2006); J. R. Sahu, C. R. Serrao, N. Ray, U. V. Waghmare, C. N. R. Rao, J. Mater. Chem. **17**, 42 (2007); W. Eerenstein, M. Wiora, J. L. Prieto, J. F. Scott, N. D. Mathur, Nature Mater. **6**, 348 (2007); W. Wu, et al., Phys. Rev. Lett. **101**, 137203 (2008); B. Kundys, A. Maignan, C. Simon, Appl. Phys. Lett. **94**, 072506, (2009); T. Kimura, Nature Mater. **7**, 291 (2008); G. Catalan, J. F. Scott, Adv. Mat. **21**, 2463 (2009); C. N. R. Rao et al., J. Phys. Chem. Lett. **3**, 2237 (2012); R. D. Johnson et al., Phys. Rev. Lett. **108**, 067201 (2012); N. Lee et al., Phys. Rev. Lett. **110**, 137203 (2013); K. Singh et al., Phys. Rev. B **88**, 094438 (2013); D. K. Pratt et al., Phys. Rev. B **90**, 140401(R) (2014); R. Saha et al., Mater. Horiz. **1**, 20 (2014); T. Basu et al., Sci. Rep. **4**, 5636 (2014); A. K. Kundu, M. M. Seikh, P. Nautiyal, J. Magn. Magn. Mater. **378**, 506, (2015)
11. C. Felser, G.H. Fecher, B. Balke, Angew. Chem. Int. Ed. **46**, 668 (2007)
12. D.J. Singh, C.H. Park, Phys. Rev. Lett. **100**, 087601 (2008)
13. M. Azuma, K. Takata, T. Saito, S. Ishiwata, Y. Shimakawa, M. Takano, J. Am. Chem. Soc. **127**, 8889 (2005)
14. I.O. Troyanchuk, O.S. Mantytskaja, H. Szymczak, M.Y. Shvedun, Low. Temp. Phys. **28**, 569 (2002)
15. Y.D. Zhao, J. Park, R.J. Jung, H.J. Noh, S.J. Oh, J. Mag. Mag. Mater. **280**, 404 (2004)
16. M. Gajek, M. Bibes, S. Fusil, K. Bouzehouane, J. Fontcuberta, A. Barthélémy, A. Fert, Nature Mater. **6**, 296 (2007)
17. C.-H. Yang, S.-H. Lee, C. Song, T.Y. Koo, Y.H. Jeong, Phys. Rev. B **75**, 140104(R) (2007)
18. A.K. Kundu, V. Pralong, V. Caignaert, C.N.R. Rao, B. Raveau, J. Mater. Chem. **17**, 3347 (2007)
19. A.K. Kundu, R. Ranjith, B. Kundys, N. Nguyen, V. Caignaert, V. Pralong, W. Prellier, B. Raveau, Appl. Phys. Lett. **93**, 052906 (2008); V.K. Jha, P. Nautiyal, M.M. Seikh, R. Chatterjee, R. Mahendiran, A.K. Kundu, J. Mater. Sci. **48**, 7629 (2013)
20. R. Ranjith, A.K. Kundu, M. Filippi, B. Kundys, W. Prellier, B. Raveau, J. Laverdiere, M. P. Singh, S. Jandl, Appl. Phys. Lett. **92**, 062909 (2008)
21. M. Filippi, B. Kundys, R. Ranjith, A.K. Kundu, W. Prellier, Appl. Phys. Lett. **92**, 212905 (2008)
22. A.K. Kundu, R. Ranjith, V. Pralong, V. Caignaert, B. Raveau, J. Mater. Chem. **18**, 4280 (2008)
23. G. Anjum, R. Kumar, S. Mollah, D.K. Shukla, S. Kumar, C.G. Lee, J. Appl. Phys. **107**, 103916 (2010)
24. A.K. Kundu, M.M. Seikh, A. Srivastava, S. Mahajan, R. Chatterjee, V. Pralong, B. Raveau, J. Appl. Phys. **110**, 073904 (2011)

25. A.K. Kundu, V.K. Jha, M.M. Seikh, R. Chatterjee, R. Mahendiran, J. Phys.: Condens. Matter. **24**, 255902 (2012)
26. K. Vijayanandhini, C. Simon, V. Pralong, Y. Bréard, V. Caignaert, B. Raveau, P. Mandal, A. Sundaresan, C. N. R. Rao, J. Phys.: Condens. Matter. **21**, 486002 (2009); M. E. Villafuerte-Castrej et al., Inorg. Chem. **50**, 8340 (2011); P. Mandal et al., RSC Advances, **2**, 292 (2012)
27. P. A. Joy, Y. B. Khollam, S. K. Date, Phys. Rev. B **62**, 8608 (2000); R. I. Dass, J. B. Goodenough, Phys. Rev. B **67**, 014401 (2003)
28. C.L. Bull, D. Gleeson, K.S. Knight, J. Phys.: Condens. Matter **15**, 4927 (2003); P. Mandal et al., Phys. Rev. B **82**, 100416 (2010); P. Mandal et al., J. Mater. Chem. **20**, 1646 (2010); S. Ghara et al., Phys. Rev. B **90**, 024413 (2014)
29. W. Prellier, A. Fouchet, B. Mercey, J. Phys.: Condens. Matter **15**, R1583 (2003)
30. D.N.H. Nam, R. Mathieu, P. Nordblad, N.V. Khiem, N.X. Phuc, Phys. Rev. B **62**, 1027 (2000)
31. A.K. Kundu, P. Nordblad, C.N.R. Rao, J. Phys.: Condens. Matter **18**, 4809 (2006)
32. K. Binder, A.P. Young, Rev. Mod. Phys. **58**, 801 (1986)
33. J.A. Mydosh, *Spin Glasses: An Experimental Introduction* (Taylor and Francis, London, 1993)
34. A.K. Kundu, P. Nordblad, C.N.R. Rao, J. Solid State Chem. **179**, 923 (2006)
35. J.B. Goodenough, A. Wold, R.J. Arnott, N. Menyuk, Phys. Rev. **124**, 373 (1961); G.H. Jonker, J. Appl. Phys. **37**, 1424 (1966)
36. N.Y. Vasanthacharya, P. Ganguly, J.B. Goodenough, C.N.R. Rao, J. Phys. C: Solid State Phys. **17**, 2745 (1984)
37. S. Hebert, C. Martin, A. Maignan, R. Retoux, M. Hervieu, N. Nguyen, B. Raveau, Phys. Rev. B **65**, 104420 (2002)
38. H.Z. Guo, A. Gupta, T.G. Calvarese, M.A. Subramanian, Appl. Phys. Lett. **89**, 262503 (2006)
39. G. Blasse, J. Phys. Chem. Solids **26**, 1969 (1965); V. L. J. Joly, P. A. Joy, S. K. Date, C. S. Gopinath, Phys. Rev. B **65**, 184416 (2002); R. I. Dass, J. Q. Yan, J. B. Goodenough, Phys. Rev. B **68**, 064415 (2003)
40. M. Sakai, A. Masuno, D. Kan, M. Hashisaka, K. Takata, M. Azuma, M. Takano, Y. Shimakawa, Appl. Phys. Lett. **90**, 072903 (2007)
41. P. Lunkenheimer, R. Fichtl, S.G. Ebbinghaus, A. Loidl, Phys. Rev. B **70**, 172102 (2004)
42. Lunkenheimer, V. Bobnar,A. V. Pronin, A. I. Ritus, A. A. Volkov, A. Loidl, Phys. Rev. B **66**, 052105 (2002)
43. F. Bordi, C. Cametti, R.H. Colby, J. Phys.: Condens. Matter **16**, R1423–R1463 (2004). references therein
44. D. Sinclair, T.B. Adams, F.D. Morrison, A.R. West, Appl. Phys. Lett. **80**, 2153 (2002)
45. T.B. Adams, D.C. Sinclair, A.R. West, Phys. Rev. B **73**, 094124 (2006)
46. A.I. Ritus, A.V. Pronin, A.A. Volkov, P. Lunkenheimer, A. Loidl, A.S. Shcheulin, A.I. Ryskin, Phys. Rev. B **65**, 165209 (2002)
47. N. Biškup, A. de Andrés, J.L. Martinez, C. Perca, Phys. Rev. B **72**, 024115 (2005)
48. R. Cabassi, F. Bolzoni, A. Gauzzi, E. Gilioli, A. Prodi, F. Licci, Phys. Rev. B **74**, 045212 (2006)
49. G. Catalan, J.F. Scott, Nature (London) **448**, E4–E5 (2007)
50. G. Catalan, Appl. Phys. Lett. **88**, 102902 (2006)
51. J.F. Scott, J. Mater. Res. **22**, 2053 (2007)
52. J.F. Scott, Phys. Rev. B **16**, 2329 (1977)
53. E. Granado, A. Garcia, J.A. Sanjurjo, C. Rettori, I. Torriani, E. Prado, R.D. Sanchez, A. Canerio, S.B. Oseroff, Phys. Rev. B **60**, 11879 (1999)
54. J. Laverdiere, S. Jandl, A.A. Mukhin, VYu. Ivanov, V.G. Ivanov, M.N. Iliev, Phys. Rev. B **73**, 214301 (2006)
55. F. Sriti, A. Maignan, C. Martin, B. Raveau, Chem. Mater. **13**, 1746 (2001)
56. N.F. Mott, *Metal-Insulator Transitions* (Taylor and Francis, London, 1990)
57. H. Chiba, T. Atou, T. Syono, J. Solid State Chem. **132**, 139 (1997)

58. A.K. Kundu, K. Ramesha, R. Seshadri, C.N.R. Rao, J. Phys.: Condens. Matter **16**, 7955 (2004). references therein
59. M.H. Cohen, J.B. Neaton, L. He, D. Vanderbilt, J. Appl. Phys. **94,** 3299 (2003); J. Liu, C. Duan, W.N. Mei, R.W. Smith, J.R. Hardy, J. Appl. Phys. **98**, 093703 (2005)
60. F.M.A. Da Costa, A.J.C. Dos Santos, Inorg Chim Acta **140**, 105 (1987)
61. S. D. Bhame, Joly V. L. Joseph, P. A. Joy, Phys Rev B **72**, 054426 (2005); K. De, R. Ray, R. N. Panda, S. Giri, H. Nakamura, T. Kohara, J. Magn. Magn. Mater. **288**, 339 (2005)
62. M.E. Lines, A.M. Glass, *Principles and Applications of Ferroelectrics and Related Materials* (Clarendon, Oxford, 1979)
63. P. Ravindran, R. Vidya, A. Kjekshus, H. Fjellvag, O. Eriksson, Phys. Rev. B **74**, 224412 (2006). references therein
64. A.K. Jonscher, J. Mater. Sci. **16**,2037 (1981); A.K. Jonscher, J. Phys. D **32**, R57 (1999)
65. R.A. Ricciardo, A.J. Hauser, F.Y. Yang, H. Kim, W. Lu, P.M. Woodward, Mater. Res. Bull. **44**, 239 (2009)
66. P. Nautiyal, M.M. Seikh, V. Pralong, A.K. Kundu, J. Magn. Magn. Mater. **347**, 111 (2013)
67. Y. Bai, Y. Xia, H. Li, L. Han, Z. Wang, X. Wu, S. Lv, X. Liu, J. Meng, J. Phys. Chem. C **116**, 16841 (2012)
68. Y. Bai, X. Liu, Y. Xia, H. Li, X. Deng, L. Han, Q. Liang, X. Wu, Z. Wang, J. Meng, Appl. Phys. Lett. **100**, 222907 (2012)

Index

A.K. Kundu, *Magnetic Perovskites*, Engineering Materials,
DOI 10.1007/978-81-322-2761-8